D0899130

INTRODUCTION TO THE THEORY OF ENTIRE FUNCTIONS

This is Volume 56 in
PURE AND APPLIED MATHEMATICS
A series of Monographs and Textbooks
Editors: PAUL A. SMITH AND SAMUEL EILENBERG
A complete list of titles in this series appears at the end of this volume

INTRODUCTION TO THE THEORY OF ENTIRE FUNCTIONS

A. S. B. HOLLAND

Department of Mathematics
University of Calgary
Calgary, Canada

ACADEMIC PRESS New York and London 1973

A Subsidiary of Harcourt Brace Jovanovich, Publishers

ACADEMIC PRESS, INC.
111 Fifth Avenue, New York, New York 10003

United Kingdom Edition published by
ACADEMIC PRESS, INC. (LONDON) LTD.
24/28 Oval Road, London NW1

Library of Congress Cataloging in Publication Data

Holland, Anthony S B
An introduction to the theory of entire functions.

(Pure and applied mathematics; a series of monographs and textbooks)
Bibliography: p.
1. Functions, Entire. I. Title. II. Series.
QA3.P8 vol. [QA353.E5] 510'.8s [515'.98] 72-88365
ISBN 0-12-352750-3

AMS (MOS) 1970 Subject Classifications: 30-01, 30-02, 30A62, 30A66

PRINTED IN THE UNITED STATES OF AMERICA

CONTENTS

Chapter III: Theorems Concerning the Modulus of a Function and Its Zeros

Chapter IV: Infinite Product Representation: Order and Type

Chapter V: Standard Functions and Characterization Theorems

Chapter VI: Functions with Real and/or Negative Zeros: Minimum Modulus I and Sequences of Functions

PREFACE

The purpose of this monograph is to acquaint the reader with some of the basic analysis and theorems central to a study of functions analytic in the entire finite complex plane. The study of these entire functions, a subclass of the family of meromorphic functions, has been assumed by some to be somewhat out of vogue in the present decade, but judging from the results of various specialist conferences, e.g., La Jolla, 1966 ["Entire Functions and Related Parts of Analysis" (*Proc. Symp. Pure Math., 11th*), AMS, 1968], this assumption is not particularly valid. This short monograph will display some of the intrinsic beauty of the subject which is far from "exhausted," and array in a sequential form theorems central to the study of entire functions. Since the chapters are developed in a logical order, for those not particularly familiar with the basic function theory it is advisable not to skip sections. An appendix has been added listing a few of the more important theorems and definitions.

An attempt has been made to prove results fully and to use the phrase "it can be shown," as seldom as possible. Anyone with a good first course in complex function theory should have little difficulty in reading all chapters.

A considerable amount of work has been accomplished, and indeed is still being done, in the subjects of distribution of zeros of polynomials, Hilbert spaces of entire functions, iteration of functions, etc. It is partly with this in mind that this survey is written, since most of what follows is basic to an understanding of not only the geometry of zeros, but the growth and behavior of the maximum modulus, the minimum modulus, the a-points, etc.

Very few of the results are new, in the sense that one could not find them in function theory texts, however the claim to originality of this survey is its continuity, its amplification of somewhat recondite theorems and its grouping of results scattered throughout a wide spectrum of texts, articles, and lecture notes. For some of the latest advances in entire function theory in particular, the reader is referred to an extensive bibliography which, by virtue of the scope of the subject, is by no means comprehensive.

I should like to add that a further purpose of this monograph is to prepare the reader for study of papers which have expanded the frontiers of the subject and, in particular, for study of such erudite works as "Meromorphic Functions," by W. K. Hayman, Oxford University Press (Clarendon), 1964, and "Entire Functions" by R. Boas, Academic Press, 1954. I have reproduced very little of the material in Dr. Boas' book, which among other things, deals at length with functions of exponential order: most of the present survey deals with functions of arbitrary order.

The difference between a lemma and a theorem is always debatable, and in this survey a lemma is used to mean a result of lesser importance than a theorem, but normally needed for the theorem which it precedes.

Proper names appended to theorems only mean that the name is that by which the theorem is normally known, and is not meant to imply that the theorem is exclusively the discovery of that particular person.

In attempting to develop a style and arrangement of material, I have looked to the works of Dr. E. C. Titchmarsh, whose several very readable texts are well known to all analysts. I have had several occasions to use material, in particular from "The Theory of Functions" by E. C. Titchmarsh, second edition, 1939, for which I wish to acknowledge the kind permission of the Oxford University Press (Clarendon).

I wish also to thank Dr. G. Pólya for permission to include his result on an entire function of an entire function, and also to acknowledge the very considerable assistance given me by my colleague, the late Dr. H. D. Ursell.

INTRODUCTION TO THE THEORY OF ENTIRE FUNCTIONS

CHAPTER I

A STUDY OF THE MAXIMUM MODULUS AND BASIC THEOREMS

We commence our study of entire functions by acquainting ourselves with some of the basic properties of analytic functions. Elementary analysis such as properties of Laurent and Taylor series expansion of functions, Cauchy's theorem, Cauchy's formulas and logarithms of complex functions has been assumed. Although there are several ways to treat the concept of an entire function, the whole treatment has been kept classical. Many far-reaching ramifications of some of the later theorems can be explored with techniques of modern function theory, for example, the use of elliptic modular functions in Picard's theorems, or the structure of Hilbert spaces of entire functions. However, the exploration of these techniques has been left to the reader who will find many references to them in the bibliography.

An initial start is now made by examining the behavior of analytic functions at singular points.

1.1 THE NATURE OF SINGULAR POINTS

A singular point of a complex function is a place where the function ceases to be analytic (see Appendix). The point z_0 is said to be an *isolated singularity* for a single-valued function $f(z)$ if $\exists \delta > 0$, such that $f(z)$ is analytic in $0 < | z - z_0 | < \delta$ but not in $| z - z_0 | < \delta$.

Most singularities we encounter are singularities of functions such as $1/(z-a)$ at $z=a$ or $1/e^z$ at $z=0$. However, a function defined to be

$$f(z) = \begin{cases} 1, & z \neq 0, \\ 0, & z = 0, \end{cases}$$

clearly has a singularity at the origin. Note that $z=0$ is not an isolated singularity of the multivalued function $f(z) = z^{1/2}$. In the first place, $z^{1/2}$ does not possess a derivative at $z=0$ and further, there is no neighborhood of $z=0$ in which the function is single-valued. Thus the requirement that $f(z)$ be single-valued excludes branch points. To analyze the behavior of a function at an isolated singularity we use a Laurent series. Let $f(z)$ have an isolated singularity at $z=z_0$. Then we know that $f(z)$ can be expanded as a Laurent series in a sufficiently small circle about z_0 (except at z_0 itself). Let the Laurent series be

$$f(z) = \sum_{n=-\infty}^{\infty} a_n(z-z_0)^n$$

(see Appendix). We divide isolated singularities into three classes.

I. Let $f(z)$ be bounded in the neighborhood of z_0. Assuming the Laurent series for $f(z)$ to exist, $\exists M > 0$, such that

$$|f(z)| \leq M, \qquad 0 < |z - z_0| \leq r.$$

Cauchy's inequalities show that $|a_n| \leq M/r^n$ for r arbitrarily small. Thus all coefficients with negative index vanish.

Hence, except for $z = z_0$, $f(z)$ is represented by a Taylor series

$$f(z) = a_0 + a_1(z - z_0) + \cdots + a_r(z - z_0)^r + \cdots.$$

The Taylor series defines an analytic function throughout the interior of a circle of convergence about z_0 and the series coincides with $f(z)$ at all of these points except $z = z_0$.

If we redefine $f(z)$ at z_0 by $f(z_0) = a_0$, then the function is analytic in $|z - z_0| \leq r$. This singularity is due to a break in the continuity of $f(z)$. Thus to make the function analytic at z_0, it suffices to redefine $f(z)$ at z_0 so as to make $f(z)$ continuous. The singularities of this type are called *removable*.

II. Let $f(z)$ be unbounded in the neighborhood of a point z_0. The Laurent expansion of f in the deleted neighborhood of z_0 contains negative index terms. Thus

$$f(z) = g(z - z_0) + h\left(\frac{1}{z - z_0}\right)$$

where

$$g(z - z_0) = \sum_{n=0}^{\infty} a_n (z - z_0)^n \qquad \text{and} \qquad h\left(\frac{1}{z - z_0}\right) = \sum_{n=1}^{\infty} a_{-n} \frac{1}{(z - z_0)^n}.$$

The series $h(1/(z - z_0))$ is called the principal part of $f(z)$. If the principal part has a finite number of terms, i.e., $a_{-n} = 0$, $n > k > 0$, and $a_{-k} \neq 0$, then

$$f(z) = \frac{a_{-k}}{(z - z_0)^k} + \frac{a_{-k+1}}{(z - z_0)^{k-1}} + \cdots$$

and we say that $f(z)$ has a *pole* of *order* k at $z = z_0$. If the principal part has an infinite number of non zero terms, we say that z_0 is an *essential singularity* of $f(z)$, e.g., $e^{1/z}$ has an essential singularity at $z = 0$.

The following theorem illustrates the nature of a function with an isolated essential singularity.

1.1.1 Theorem (Weierstrass). In the neighborhood of an isolated essential singularity, an analytic function comes arbitrarily close to every complex value. (Picard's theorem states that in every neighborhood of an essential singularity, an analytic function takes on every finite complex value with one possible exception).

Proof. Let $f(z)$ have an essential singularity at z_0. Let $f(z)$ be bounded away from the value a in a neighborhood of z_0, i.e., let $|f(z) - a|$ be bounded away from zero. Then

$$g(z) = \frac{1}{f(z) - a}$$

is bounded and analytic in a deleted neighborhood of z_0, i.e.,

$$0 < |z - z_0| < R.$$

Thus $g(z)$ has at most a removable singularity at z_0. Further, $g(z_0) \neq 0$, for otherwise

$$f(z) = a + \frac{1}{g(z)}$$

would have only a pole at z_0. Thus if $g(z)$ approaches a nonzero limit as

$z \to z_0$, then $f(z)$ has a removable singularity at z_0, which conflicts with the assumption. Thus $f(z)$ is not bounded away from a value a in the neighborhood of z_0. □

We call a point of accumulation of poles in a domain where $f(z)$ is otherwise analytic an *essential singularity*.

III. Let z_0 be an essential singularity of $f(z)$. Let $f(z)$ be bounded below in the neighborhood of z_0. That is, $\exists M > 0$ such that

$$|f(z)| \geq M \qquad \text{for} \quad 0 < |z - z_0| < \delta.$$

Then $1/f(z)$ is analytic and z_0 is a point of accumulation of zeros of $1/f(z)$ in its domain of analyticity. Hence $1/f(z)$ is identically zero which is clearly impossible. Thus an accumulation point z_0 of poles does not behave like a pole.

Singularities at infinity may be discussed similarly, since if $f(z)$ is analytic in

$$|z| > R, 0 \leq R < \infty,$$

then $F(\omega) \equiv f(1/\omega)$ is analytic for

$$0 < |\omega| < 1/R$$

and the three cases follow. Further, an infinity of separate poles can have an accumulation point at infinity.

1.2 Definition. A *meromorphic function* has no singularities in the finite plane except poles. A meromorphic function is *rational meromorphic* if $z = \infty$ is a regular point or a pole and *transcendental meromorphic* if $z = \infty$ is an essential singularity.

One simple characterization of a meromorphic function is as follows.

1.2.1 Theorem. A rational meromorphic function must be a rational function (i.e., quotient of two polynomials).

Proof. Let $f(z)$ be a rational meromorphic function. Let the points z_0, $z_1, \ldots, z_n$ (one of which may be the point at ∞) include all the poles. The number of poles is finite, otherwise they would possess a point of accumulation in the whole plane and the function would possess an essential singularity.

Let

$$p_0\left(\frac{1}{z - z_0}\right), \ldots, p_n\left(\frac{1}{z - z_n}\right)$$

denote the principal parts of the Laurent expansion of $f(z)$ about $z_0, \ldots, z_n$, respectively. Then

$$f(z) - \sum_{\nu=0}^{n} p_\nu\left(\frac{1}{z - z_\nu}\right)$$

is analytic in the whole z-plane including the point at ∞. Thus it is uniformly bounded and by Liouville's theorem it may be shown to be constant. Hence

$$f(z) = \sum_{\nu=0}^{n} p_\nu\left(\frac{1}{z - z_\nu}\right) + C.$$

However, the p_ν's are polynomials in the respective arguments, therefore $f(z)$ is the quotient of two polynomials, i.e.,

$$f(z) \equiv \frac{P(z)}{Q(z)},$$

and the equation gives the decomposition of $f(z)$ into its partial fractions where the z_ν are the zeros of $Q(z)$. □

1.3 Definition. An *entire function* is a function analytic in the entire finite complex plane.

Thus an entire function may be represented by an everywhere convergent power series $\sum_{n=0}^{\infty} a_n z^n$. Entire functions are special cases of analytic functions.

There are three ways for an entire function to behave at infinity:

i. $f(z)$ can have a regular point at ∞. Then we will show by Liouville's theorem $f(z) \equiv K$.

ii. $f(z)$ can have a pole of order $k \geq 1$ at ∞. Then $f(z)$ is a polynomial.

iii. $f(z)$ can have an essential singular point at ∞. Then $f(z)$ is a transcendental entire function.

The sum, difference and product of a finite number of entire functions are entire functions, and the quotient of two entire functions is an entire function (provided that the denominator is nowhere zero). An entire function of an entire function is an entire function.

We consider the behavior of one of the simplest of entire functions e^z.

Since

$$e^z \cdot e^{-z} = 1 \qquad \text{for all finite } z,$$

e^z has no zeros in the finite plane. Since

$$e^z = A \neq 0 = e^{|\log A| + i(\arg A + 2n\pi)}, \qquad n = 0, \pm 1, \ldots,$$

then

$$1 + z + \frac{z^2}{2!} + \cdots + \frac{z^n}{n!} + \cdots = A \neq 0$$

has infinitely many roots.

NOTE. $\log z$ is not an entire function (see $z = 0$, for example). However, since

$$\begin{aligned} \log e^z &= \log |e^z| + i \arg e^z \\ &= \log e^x + i(y + 2n\pi) = x + iy + i2n\pi = z + 2n\pi i, \end{aligned}$$

$\log e^z$ represents an infinite family of entire functions, viz. z, $z + 2\pi i$, $\ldots$.

1.3.1 Theorem. If $f(z)$ is an entire function which does not vanish at any point, then $f(z) = e^{g(z)}$ where $g(z)$ is an entire function.

Proof. Since $\log f(z)$ represents an infinite family of entire functions differing from each other by integral multiples of $2\pi i$, denote any one of these by $g(z)$ and we have that

$$\log f(z) = g(z) + 2n\pi i.$$

Thus

$$f(z) = e^{\log f(z)} = e^{g(z) + 2n\pi i} = e^{g(z)}.$$

($2\pi i$ is the period of the exponential function.) □

1.4 MAXIMUM MODULUS

We now study the maximum modulus $M(r) = \max |f(z)|$ of a function $f(z)$ analytic on the disk $|z| \leq r$. The maximum principle illustrated by the maximum modulus theorem, a theorem central to all analytic function theory, demonstrates the remarkable property of these functions, viz., that if a function $f(z)$ is analytic in an open connected set K and is not constant in K, then $f(z)$ has no maximum value in K. Further, if $f(z)$ is continuous in a closed connected set D and analytic in the interior of D, then $|f(z)|$ attains its maximum value on the boundary of D.

The problem of finding the exact position where a function attains its maximum absolute value can offer considerable difficulty. We expect the maximum modulus of an entire function to be on the boundary of the domain D since (as we shall show in Section 6.3) $M(r)$ increases with r, however, this is certainly not evident for all analytic functions. We prove the maximum modulus theorem for a function analytic in and continuous on the boundary of a domain D.

Consider $f(z)$ to be analytic in a domain D, including the origin. Write

$$f(z) = \sum_{n=0}^{\infty} a_n z^n, \qquad |z| < R.$$

For $r < R$,

$$|f(z)|^2 = f(z) \cdot \overline{f(z)} = \sum_{m=0}^{\infty} a_m r^m e^{im\theta} \cdot \sum_{n=0}^{\infty} \bar{a}_n r^n e^{-in\theta}.$$

Since both series are absolutely convergent (within the original circle of convergence), they may be multiplied, the resulting series being uniformly convergent for $0 \leq \theta \leq 2\pi$ and for any given $r < R$ ($|z|$ in general cannot approach R).

1.4.1 Lemma. If $f(z)$ is analytic for $|z - a| < R$, $R > 0$, write

$$M(r) = \max_{|z-a|=r} |f(z)|, \qquad 0 \leq r < R.$$

Then

$$|f(a)| \leq M(r), \quad 0 \leq r < R, \qquad \text{and} \qquad |f(z)| = M(r), \quad 0 < r < R$$

if and only if $f(z) = \text{constant} = M(r)e^{i\alpha}$ (α real) and $|z - a| < R$.

Proof. Since we may write

$$f(z) = \sum_{n=0}^{\infty} a_n (z-a)^n, \qquad |z - a| < R,$$

we have

$$\sum_{n=0}^{\infty} |a_n|^2 r^{2n} = \frac{1}{2\pi} \int_0^{2\pi} |f(a + re^{i\theta})|^2 \, d\theta, \qquad 0 \leq r < R$$

[valid at $r = 0$ since $|a_0|^2 = |f(a)|^2$]. Thus

$$\sum_{n=0}^{\infty} |a_n|^2 r^{2n} \leq \{M(r)\}^2$$

and $|a_0|^2 \leq \{M(r)\}^2$, i.e., $|a_0| \leq M(r)$. Also, since

$$|a_0|^2 + |a_1|^2 r^2 + \cdots + |a_n|^2 r^{2n} \leq \{M(r)\}^2$$

$|a_0| < M(r)$, $0 < r < R$, unless $a_n = 0$, $n = 1, 2, \ldots$ whence $f(z)$ is a constant $= M(r)$, $0 < r < R$. Thus either $f(z)$ is constant for $|z - a| < R$ or $|f(z)|$ takes values greater than $|f(a)|$ in every neighborhood of $z = a$.

Alternatively, if $f(z)$ is analytic and not constant in a domain D, then $|f(z)|$ cannot take a maximum value at any point $a \in D$. □

1.4.2 The Maximum Modulus Theorem. If $f(z)$ is analytic in a bounded domain D and continuous on $\bar{D}$ and $M = \max_{z \in C} |f(z)|$, where C is the boundary of D, then

$$|f(z)| \leq M \qquad \text{in } D.$$

Further, $|f(z)| < M$ in D except in the case where $f(z)$ is a constant.

Proof. a. Clearly, if $f(z) = K$ (constant) $\forall z$, $z \in D$, then since f is continuous on $\bar{D}$, $f(z) = K \ \forall z$, $z \in \bar{D}$, and hence $f(z) = K \ \forall z$, $z \in C$ so that $|K| = M$ and $|f(z)| = M \ \forall z$, $z \in \bar{D}$.

b. Assume $f(z)$ is not constant in D. Write

$$K = \max_{z \in \bar{D}} |f(z)|.$$

Since $\bar{D}$ is compact, K is finite and $|f(z)|$ attains the value K at least once in $\bar{D}$. By the previous lemma,

$$|f(z)| \neq K, \qquad \forall z = a \in D$$

and $|f(z)| > |f(a)|$ in every neighborhood of a. Thus the maximum value K of $|f(z)|$ on $\bar{D}$, is attained at some $z \in C$. Accordingly $K = M$ and $|f(z)| < M$ in D. □

1.5 The Minimum Principle. If $f(z)$ is analytic in D and is never zero there, then $[f(z)]^{-1}$ is analytic in D.

Since the minimum of $|f(z)|$, i.e., $m(r)$, is attained at the same points as the maximum of $|f(z)|^{-1}$, it follows from the maximum principle that $|f(z)|$ cannot attain its minimum in the interior of D. Further, since the inability of a function to attain its maximum at an interior point is a "local" property, the maximum principle is valid for analytic functions not single-valued in a multiconnected domain.

Thus we have the minimum principle, viz., that $|f(z)|$ attains its minimum value on the boundary of a region D and that if its minimum value is found inside D, then f is necessarily a constant.

With the maximum modulus theorem at our disposal, we now consider $M(r)$ first from the point of view of its relation to the coefficients in the power series for $f(z)$, later in its growth as compared to the modulus of a standard function like e^z. Later still, we compare the growth of the maximum modulus to the growth of the minimum modulus.

It transpires that in many cases the growth of the maximum modulus is not a particularly suitable object with which to study the more delicate behavior of entire functions. A new object called the *Nevanlinna characteristic* is introduced to which a final section of the survey will be devoted.

1.6 MAXIMUM MODULUS OF COEFFICIENTS OF A POWER SERIES

Let

$$f(z) = a_0 + a_1 z + \cdots + a_p z^p + \cdots, \qquad |z| \le r < R,$$

and hence

$$f(z)/z^p = a_0 z^{-p} + a_1 z^{-p+1} + \cdots + a_p + \cdots.$$

Integrating with respect to ϕ from 0 to 2π,

$$\int_0^{2\pi} f(z)/z^p \, d\phi = \int_0^{2\pi} a_0 z^{-p} \, d\phi + \cdots + \int_0^{2\pi} a_p \, d\phi + \cdots.$$

(We can integrate the infinite series, since for variable ϕ and fixed r the series converges uniformly.) Now

$$\int_0^{2\pi} z^m \, d\phi = \int_0^{2\pi} r^m(\cos m\phi + i \sin m\phi) \, d\phi = r^m \left[\frac{\sin m\phi}{m} - i \frac{\cos m\phi}{m} \right]_0^{2\pi} = 0.$$

Thus $2\pi a_p = \int_0^{2\pi} f(z)/z^p \, d\phi$ and

$$a_p = \frac{1}{2\pi} \int_0^{2\pi} f(z)/z^p \, d\phi \qquad \text{for} \quad p = 0, 1, 2, \ldots.$$

Let $M(r)$ denote the maximum modulus of $f(z)$ on a closed disk of radius r, center $z = 0$, i.e., $M(r) = \max_{|z| \le r} |f(z)|$. Since

$$|a_p| \le \frac{1}{2\pi} \int_0^{2\pi} \frac{|f(z)|}{|z|^p} \, d\phi$$

with $|z| = r$, $|f(z)| \leq M(r)$, we have

$$|a_p| \leq M(r)/r^p \qquad \text{(Cauchy's inequality).}$$

Three theorems follow immediately.

1.6.1 Theorem. A function which is analytic and bounded for all finite z is constant.

Proof. If $f(z)$ is analytic for all finite z, the Taylor series $f(z) = \sum_{n=0}^{\infty} a_n z^n$ converges for all z. If $|f(z)| \leq M$, we have $|a_n| \leq Mr^{-n}$ for all n and r. Letting $r \to \infty$, $|a_n| \leq 0$ for $n > 0$. Thus $a_n = 0, n > 0$, and $f(z) = a_0$. □

1.6.2 Theorem. If $f(z)$ is analytic for all finite z and if as $|z| \to \infty$

$$|f(z)| = O(|z|^k),$$

then $f(z)$ is a polynomial of degree less than or equal to k.

Proof. By Cauchy's inequality $|a_n| \leq M(r)r^{-n}$. Since $M(r) = O(r^k)$

$$|a_n| \leq O(r^{k-n}) \to 0, \qquad \text{for} \quad n > k, \quad r \to \infty$$

and

$$a_n = 0 \qquad \text{for} \quad n > k.$$

Thus $f(z)$ is a polynomial of degree less than or equal to k. □

1.6.3 Theorem. (Liouville). If an entire function is not an absolute constant, then $M(r) \to \infty$, $r \to \infty$.

Proof. Let $f(z) = a_0 + a_1 z + \cdots a_n z^n + \cdots$ be an entire function. Since $|a_n| \leq M(r)/r^n$ for $M(r)$ the maximum modulus of $f(z)$ on $|z| \leq r$ and since $M(r)$ is a nondecreasing function of r, either $\exists$ a constant C such that

$$M(r) \leq C, \qquad C > 0,$$

or

$$M(r) \to \infty, \qquad r \to \infty.$$

Assume the existence of C, then

$$|a_n| \leq C/r^n \qquad \text{for} \quad r > 0, \quad n = 0, 1, \ldots .$$

For $n \geq 1$, $C/r^n \to 0$, $r \to \infty$. Since the left-hand side is independent of r, we have $|a_n| \leq 0$, i.e., $a_n = 0$, $n \geq 1$, and thus $f(z) \equiv a_0$. If $f(z) \not\equiv$ constant, then $M(r)$ is not bounded, and since it is nondecreasing, $M(r) \to \infty$, $r \to \infty$. □

1.7 MAXIMUM MODULUS OF A POLYNOMIAL

Consider a polynomial

$$P(z) = a_0 + a_1 z + \cdots + a_n z^n, \qquad a_n \neq 0, \quad n \geq 1.$$

Then

$$\begin{aligned} |P(z)| &\leq |a_0| + |a_1| |z| + \cdots + |a_n| |z|^n \\ &\leq |a_0| + |a_1| r + \cdots + |a_n| r^n, \qquad |z| \leq r \end{aligned}$$

and

$$|P(z)| \leq |a_n| r^n \left\{ 1 + \left(\frac{|a_{n-1}|}{|a_n|} \frac{1}{r} + \cdots + \frac{|a_0|}{|a_n|} \frac{1}{r^n} \right) \right\}.$$

Clearly, as $r \to \infty$

$$\left(\frac{|a_{n-1}|}{|a_n|} \frac{1}{r} + \cdots \right) \to 0.$$

Thus for $\varepsilon > 0$, in particular, $\varepsilon < 1$, $\exists r_0(\varepsilon)$ such that for $r > r_0(\varepsilon)$,

$$\frac{|a_{n-1}|}{|a_n|} \frac{1}{r} + \cdots + \frac{|a_0|}{|a_n|} \frac{1}{r^n} < \varepsilon.$$

Hence for $r > r_0(\varepsilon)$ and $|z| \leq r$,

$$|P(z)| \leq |a_n| r^n (1 + \varepsilon).$$

Similarly, for a point z_0 on $|z| = r$

$$\begin{aligned} |P(z)| &= |a_n z_0^n + (a_{n-1} z_0^{n-1} + \cdots + a_0)| \\ &\geq |a_n| \, |z_0|^n - |a_{n-1} z_0^{n-1} + \cdots + a_0| \\ &\geq |a_n| \, |z_0|^n - |a_{n-1}| \, |z_0|^{n-1} - \cdots - |a_0| \\ &= |a_n| r^n - \{ |a_{n-1}| r^{n-1} + \cdots + |a_0| \} \\ &= |a_n| r^n \left\{ 1 - \left(\frac{|a_{n-1}|}{|a_n|} \frac{1}{r} + \cdots + \frac{|a_0|}{|a_n|} \frac{1}{r^n} \right) \right\} \end{aligned}$$

and for $r > r_0(\varepsilon)$

$$|P(z_0)| \geq |a_n| r^n(1 - \varepsilon).$$

Thus for arbitrary $\varepsilon > 0$ and sufficiently large r

$$|a_n| r^n(1 - \varepsilon) \leq |P(z)| \leq |a_n| r^n(1 + \varepsilon).$$

This inequality is true at a point in the disk $|z| \leq r$ at which $|P(z)|$ attains its maximum $M(r)$, i.e.,

$$M(r) \leq |a_n| r^n(1 + \varepsilon), \qquad r > r_0(\varepsilon).$$

Also since the value of $|P(z)|$ at a point z_0 on $|z| = r$ cannot exceed $M(r)$, thus $M(r) \geq |a_n| r^n(1 - \varepsilon)$ and

$$1 - \varepsilon \leq M(r)/|a_n| r^n \leq 1 + \varepsilon.$$

Since ε is arbitrarily small, we have

$$\lim_{r \to \infty} M(r)/|a_n| r^n = 1,$$

i.e., the maximum modulus of a polynomial of degree n is asymptotically equal to the modulus of the highest degree term of the polynomial.

1.8 MAXIMUM MODULUS OF AN ENTIRE FUNCTION

We define $M(r; f)$ to be the maximum modulus of $f(z)$ in a closed disk $|z| \leq r$, e.g.,

$$M(r; \cos z) = \max_{|z| \leq r} |\cos z|.$$

Since

$$\cos z = 1 - \frac{z^2}{2!} + \cdots, \qquad |z| < \infty,$$

then

$$|\cos z| \leq 1 + \frac{|z|^2}{2!} + \frac{|z|^4}{4!} + \cdots.$$

Hence

$$|\cos z| \leq 1 + \frac{r^2}{2!} + \frac{r^4}{4!} + \cdots = \cosh r$$

and

$$|\cos z| \leq \frac{e^r + e^{-r}}{2}, \qquad |z| \leq r.$$

However, at $z = ir$ on the circle,

$$\cos z = \cos ir = \frac{e^{i(ir)} + e^{-i(ir)}}{2} = \frac{e^{-r} + e^{r}}{2}$$

and $(e^r + e^{-r})/2$ coincides with the maximum modulus of $\cos z$ in $|z| \leq r$. Thus

$$M(r;\ \cos z) = (e^r + e^{-r})/2.$$

Similarly

$$M(r;\ \sin z) = (e^r - e^{-r})/2 \qquad \text{and} \qquad M(r;\ e^z) = e^r.$$

1.8.1 Theorem. If an entire function is not a polynomial, its maximum modulus increases faster than the maximum modulus of any polynomial, i.e.,

$$\lim_{r\to\infty} M(r;\ P)/M(r; f) = 0,$$

where $P(z)$ is the polynomial and $f(z)$ is the entire function.

Proof. Write $f(z) = a_0 + a_1 z + \cdots + a_n z^n + \cdots$, and let $P(z)$ be of degree m. Denote the coefficient of the highest degree term of the polynomial by $b_m \neq 0$ and for $\varepsilon_0 > 0$,

$$M(r;\ P) \leq b_m r^m (1 + \varepsilon_0), \qquad r > r_0.$$

Consider $f(z)$: choose a term of degree $p \geq m + 1$ $(a_p \neq 0)$. From Cauchy's inequality, $|a_p| \leq M(r;\ f)/r^p$ and we have

$$M(r;\ f) \geq r^p |a_p| \geq |a_p| r^{m+1}.$$

Choose $r > 1$, then for $r > r_0$,

$$\frac{M(r; P)}{M(r; f)} \leq \frac{|b_m| r^m (1 + \varepsilon_0)}{|a_p| r^{m+1}} = \frac{|b_m| (1 + \varepsilon_0)}{|a_p| r}.$$

Thus $\lim_{r\to\infty} M(r;\ P)/M(r;\ f) = 0$. ☐

1.9 ORDER OF THE ZEROS OF A NONCONSTANT ENTIRE FUNCTION

If $f(z) = a_0 + a_1 z + \cdots + a_n z^n + \cdots$, we expand about $z = a$ to obtain

$$f(z) = c_0 + c_1(z - a) + \cdots + c_n(z - a)^n + \cdots.$$

If a is a zero of $f(z)$, then $c_0 = 0$. It may be that c_1, c_2, etc. are also zero. (Not all $c_i = 0$ since we would have $f(z) \equiv$ constant.)

Thus $\exists k$, the smallest number for which $c_k \neq 0$. The rank k of this coefficient or exponent in the expression $(z-a)^k$ is called the *order* of the zero. Thus if the order of a zero a of $f(z)$ is k,

$$f(z) = c_k(z-a)^k + c_{k+1}(z-a)^{k+1} + \cdots, \qquad c_k \neq 0, \quad k \geq 1,$$

and

$$f(z) = (z-a)^k[c_k + c_{k+1}(z-a) + \cdots].$$

Note the series converges for arbitrary z. The series represents an entire function and $f(z) = (z-a)^k\phi(z)$, where $\phi(z)$ is an entire function of which $z = a$ is not a zero.

1.10 Definition. An analytic function $f(z)$ is said to be *algebraic* if it satisfies an equation

$$P_0(z) + P_1(z)f(z) + P_2(z)f^2(z) + \cdots + P_n(z)f^n(z) = 0,$$

for all z in some given domain, and where $P_i(z)$ are polynomials (of arbitrary finite degree) for $i = 0, \ldots, n$, $n \geq 1$, and $P_n(z) \not\equiv 0$.

We recall the definition of a transcendental number to be a (real) number which is not a root of any algebraic equation with rational coefficients and of finite degree.

1.10.1 Theorem. If an entire function is not a polynomial, it is nonalgebraic.

Proof. Let $f(z)$ be a transcendental entire function, and suppose that it satisfies the polynomial equation of degree n,

$$P_0(z) + \cdots + P_n(z)f^n(z) = 0.$$

Consider a sequence of disks, centered at the origin, radius $r = 1, 2, 3, \ldots$. Let z_k be a point in the disk of radius k, at which $|f(z)|$ attains a local maximum, i.e.,

$$|f(z_k)| = M(k; f).$$

Since $\lim_{k\to\infty} M(k; f) = \infty$, then $|f(z_k)|$ is unbounded as $k \to \infty$ and thus we may assume $f(z_k) \neq 0$, $k \geq 1$. Set $z = z_k$ and thus

$$P_n(z_k) + \frac{P_{n-1}(z_k)}{f(z_k)} + \cdots + \frac{P_0(z_k)}{f^n(z_k)} = 0.$$

If $P_n(z_k)$ is a polynomial of degree greater than or equal to one,

$$\lim_{k\to\infty} | P_n(z_k) | = \infty \qquad (\text{since} \quad | z_k | \to \infty \quad \text{as} \quad k \to \infty).$$

Thus we have that

$$\left| \frac{P_{n-1}(z_k)}{f(z_k)} \right| \leq \frac{M(k; P_{n-1})}{M(k; f)}$$

and

$$\frac{| P_{n-m}(z_k) |}{| f(z_k) |^m} \leq \frac{M(k; P_{n-m})}{M(k; f)} \cdot \frac{1}{\{M(k; f)\}^{m-1}}, \qquad m \geq 1.$$

Since $\lim_{k\to\infty} M(k; f) = \infty$, the right-hand side tends to zero. Thus

$$\frac{| P_{n-m}(z_k) |}{| f(z_k) |} \to 0, \qquad m \geq 1,$$

implying $P_n(z_k) \to 0$. However $P_n(z_k)$ cannot approach zero since P_n is either a polynomial of degree greater than or equal to one or a nonzero constant. Therefore, by contradiction f is nonalgebraic. □

We now touch on one of the methods by which we characterize an entire function, that is, the "order," which compares the rate of growth of the maximum modulus with the growth of the modulus of a simple entire function, namely e^z. Evgrafov† devotes much space to this concept.

Several definitions of order, all of which are equivalent, will be presented as they are required.

1.11 RATE OF INCREASE OF THE MAXIMUM MODULUS

Since

1. $M(r; e^z) = e^r$,
2. $M(r; \exp z^k) = \exp r^k$,
3. $M(r; \exp e^z) = \exp e^r$,

then

$$\log M(r; e^z) = r, \qquad \log\log (M(r; e^z) = \log r,$$

$$\log M(r; \exp z^k) = r^k, \qquad \log\log M(r; \exp z^k) = k \log r,$$

$$\log M(r; \exp e^z) = e^r, \qquad \log\log M(r; \exp e^z) = r.$$

† M. A. Evgrafov, "Asymptotic Estimates and Entire Functions." Gordon and Breach, New York, 1961.

These observations prompt us to formulate the following definition of the growth of an entire function.

1.11.1 Definition. The order ϱ of an entire function $f(z)$ not identically zero, is defined as the superior limit of the ratio of $\log\log M(r; f)$ to $\log\log M(r; e^z)$ as $r \to \infty$, i.e.,

$$\varrho = \overline{\lim_{r\to\infty}} \frac{\log\log M(r; f)}{\log\log M(r; e^z)} = \overline{\lim_{r\to\infty}} \frac{\log\log M(r; f)}{\log r},$$

e.g., if $f(z) = \cos z$

$$M(r;\ \cos z) = \frac{e^r + e^{-r}}{2} = e^r \cdot \frac{1 + e^{-2r}}{2}$$

and

$$\frac{\log\log M(r; \cos z)}{\log r} = \frac{\log \left| r\left[1 + r^{-1}\left(\log \frac{1 + e^{-2r}}{2}\right)\right]\right|}{\log r},$$

therefore

$$\varrho = \overline{\lim_{r\to\infty}} \left[1 + \frac{\log\left[1 + r^{-1}\left(\log \frac{1 + e^{-2r}}{2}\right)\right]}{\log r}\right] = 1.$$

For arbitrary entire functions, order is somewhat difficult to calculate exactly. However, more often than not an estimate of the magnitude of this expression is good enough.

1.12 Theorem. If $f(z)$ is an entire function not identically zero, then for all $z = z_0$ in the plane, $\exists$ a disk centered at z_0, in which $f(z)$ has no zeros except possibly $z = z_0$ itself.

Proof. Suppose $f(z_0) \neq 0$. Then $|f(z_0)|$ is a positive number. Since $f(z)$ is continuous (an entire function, therefore differentiable), $\exists$ a disk centered at z_0, such that $|f(z) - f(z_0)| < \varepsilon$ for all $\varepsilon > 0$. Thus

$$\begin{aligned} |f(z)| &= |f(z_0) + [f(z) - f(z_0)]| \\ &\geq |f(z_0)| - |f(z) - f(z_0)| \\ &> |f(z_0)| - \varepsilon. \end{aligned}$$

Taking $\varepsilon = |f(z_0)|$, we have

$$|f(z)| > |f(z_0)| - |f(z_0)| = 0$$

and hence $|f(z)| \neq 0$. Thus for $f(z_0) \neq 0$, $\exists$ a disk centered at z_0 containing no zero of $f(z)$. Further, suppose $f(z_0) = 0$ and k is the order of the zero $z = z_0$. Then $f(z) = (z - z_0)^k \phi(z)$, where $\phi(z)$ is an entire function and $\phi(z_0) \neq 0$. We have just proved $\exists$ a disk centered at z_0, in which $\phi(z) \neq 0$, hence in that disk $f(z)$ has no zeros other than z_0. □

For the next few theorems we need one or two fundamental properties of the complex number system. We assume the following property.

1.13 The Nested Interval Property. Given a nested sequence of closed (bounded) regions in the complex plane, viz., $I_1, I_2, I_3, \ldots$, such that $I_{n+1} \subseteq I_n$ $(n = 1, 2, \ldots)$, then $\exists$ at least one point common to all the regions. And we prove the following theorem.

1.13.1 The Bolzano–Weierstrass Theorem. Every bounded infinite set S in R^2 or C has a point of accumulation.

Proof. We must find a point P such that every neighborhood of P contains infinitely many points of S. Since S is bounded, we may assume that it lies completely in a square K_0 of side l. We then divide K_0 into four equal squares by intersecting lines. One of these smaller squares must contain infinitely many points of S, for otherwise S would be finite. Denote such a square by K_1 and divide K_1 into four squares obtaining K_2 that contains infinitely many points of S. We repeat the process continuously, obtaining a sequence of squares (all closed and bounded) with

$$K_0 \supset K_1 \supset K_2 \supset \cdots \supset K_n \supset \cdots$$

each of which contains infinitely many points of S. By the nested set property there must be a point P that lies in all K_n. We want to show that this point P is a point of accumulation of S.

Let U be any neighborhood of P, a disk with center P and radius ε. We note from our construction that the length of the sides of K_n is $l/2^n$. Hence when n is sufficiently large, such that $l/2^n < \varepsilon/2$, the square K_n must lie completely in U. Hence U contains infinitely many points of S. □

1.13.2 Theorem. An entire function $f(z)$ not identically zero cannot have infinitely many zeros in a disk of finite radius.

Proof. Suppose $f(z)$ has infinitely many zeros in the disk $|z| \leq r$. Then by the Bolzano–Weierstrass theorem there exists a point z_0 on the disk which is a point of accumulation of the set of zeros of $f(z)$. Therefore in an arbitrary disk centered at z_0, there are infinitely many zeros of $f(z)$ contradicting Theorem 1.12. □

NOTE. An entire function may have an infinite set of zeros in the whole complex plane, e.g.,

$$\sin z = 0 \quad \text{for} \quad z = n\pi, \quad n = 0, \pm 1, \ldots .$$

1.13.3 Theorem. If the values of two entire functions $f(z)$ and $g(z)$ coincide at infinitely many points belonging to a disk K of finite radius, then the functions are identically equal, i.e., $f(z) \equiv g(z)$.

Proof. Since f and g are entire functions, the difference is an entire function, i.e.,

$$f(z) - g(z) = \phi(z),$$

where $\phi(z)$ vanishes at all points for which $f(z) = g(z)$. If $\phi(z) \not\equiv 0$, we have a contradiction from the previous theorem. Thus $\phi(z) \equiv 0$, i.e., $f(z) = g(z)$ $\forall z$. In particular, if the entire function $f(z)$ takes the same value A at an infinite number of points on a disk K, then $f(z) \equiv A$. □

1.13.4 Theorem. The equation

$$a_0 + a_1 z + \cdots + a_n z^n = 0, \qquad n \geq 1, \quad a_n \neq 0,$$

has at least one complex root.

Proof. (By contradiction). Let

$$P_n(z) = a_0 + a_1 z + \cdots + a_n z^n$$

and assume $P_n(z) \neq 0$ in the complex plane, i.e., $f(z) = 1/P_n(z)$ is an entire function (quotient of two entire functions). Also, $f(z)$ is nonconstant since $P_n(z) \to \infty$ as $z \to \infty$. We have then that since f is a nonconstant entire function

$$\lim_{r \to \infty} M(r; f) = \infty.$$

However, $f \to 0$ as $|z| \to \infty$ and the contradiction proves the theorem. □

We have the immediate corollary that a polynomial of degree n has exactly n roots by writing

$$P_n(z) = a_0 + a_1 z + \cdots + a_n z^n = (z - z_1) P_{n-1}(z)$$

where z_1 is the (complex) root whose existence was established and where $P_{n-1}(z)$ is a polynomial of degree $n - 1$.

1.14 We can regard a transcendental entire function

$$f(z) = a_0 + a_1 z + \cdots + a_n z^n + \cdots$$

as a sort of polynomial of infinite degree. If the analogy is valid,

$$a_0 + a_1 z + \cdots + a_n z^n + \cdots = 0$$

must have infinitely many solutions. Clearly this is not true generally. For example,

$$1 + \frac{z}{1!} + \frac{z^2}{2!} + \cdots + \frac{z^n}{n!} + \cdots = 0$$

or $e^z = 0$ has no solutions for finite z. To improve the analogy, consider

$$a_0 + a_1 z + \cdots + a_n z^n = A,$$

where A is an arbitrary complex number. This polynomial equation of degree n has n roots. Thus with $P_n(z) = A$, where $P_n(z)$ is a polynomial of degree n having roots equal in number to the degree, let us consider, for example,

$$1 + \frac{z}{1!} + \frac{z^2}{2!} + \cdots = A$$

which has infinitely many roots for $A \neq 0$. Thus the number of roots in a sense equals the degree. Similarly, by writing $\cos z = A$, $A \neq 0$, the equation has infinitely many solutions which we see by writing

$$\cos z = \frac{e^{iz} + e^{-iz}}{2}, \qquad \text{etc.}$$

This concept will be generalized later when we discuss Picard-like theorems.

1.15 Hadamard's Three-Circle Theorem. If $f(z)$ is analytic and single-valued in $\varrho < |z| < R$ and continuous on $|z| = \varrho$, $|z| = R$ and if $M(r)$ denotes the maximum of $|f(z)|$ on $|z| = r$, $\varrho < r < R$, then $\log M(r)$ is a convex function of $\log r$, i.e., for $\varrho < r_1 < r_2 < r_3 < R$,

$$\log M(r_2) \leq \log M(r_1) \frac{\log r_3 - \log r_2}{\log r_3 - \log r_1} + \log M(r_3) \frac{\log r_2 - \log r_1}{\log r_3 - \log r_1} \tag{1}$$

[We recall that a function is convex (downward) in the following sense namely, $y = \phi(x)$ is *convex* if the curve $\phi(x)$ between x_1, x_2 is always below the chord joining $(x_1, \phi(x_1))$ $(x_2, \phi(x_2))$, i.e.,

$$\phi(x) < \frac{x_2 - x}{x_2 - x_1} \phi(x_1) + \frac{x - x_1}{x_2 - x_1} \phi(x_2).]$$

Equation (1) is equivalent to

$$M(r_2)^{\log\{r_3/r_1\}} \leq M(r_1)^{\log\{r_3/r_2\}} M(r_3)^{\log\{r_2/r_1\}}.$$

Proof. Let $\phi(z) = z^\lambda f(z)$, where λ is to be a *determined constant* (real). Since the function $z^\lambda f(z)$ is not in general single-valued in $r_1 \leq |z| \leq r_3$, we cut the annulus along the negative part of the real axis obtaining a domain in which the principal branch of this function is analytic. The maximum modulus of this branch of the function in the cut annulus is attained on the boundary of the domain. Since λ is real, all the branches of $z^\lambda f(z)$ have the same modulus. By considering a branch of the function analytic in that part of the annulus for which $\pi/2 \leq \arg z \leq 3\pi/2$ it is clear that the principal value cannot attain its maximum modulus on the cut and therefore must attain it on one of the boundary circles of the annulus. Thus the maximum of $|\phi(z)|$ occurs on one of the bounding circles, i.e.,

$$|\phi(z)| \leq \max\{r_1{}^\lambda M(r_1),\ r_3{}^\lambda M(r_3)\}. \tag{1}$$

Hence on $|z| = r_2$,

$$|f(z)| \leq \max\{r_1{}^\lambda r_2^{-\lambda} M(r_1),\ r_3{}^\lambda r_2^{-\lambda} M(r_3)\} \tag{2}$$

We choose λ such that $r_1{}^\lambda M(r_1) = r_3{}^\lambda M(r_3)$ and

$$\lambda = -\{\log M(r_3)/M(r_1)\}/\{\log r_3/r_1\}.$$

With this λ in (2),

$$M(r_2) \leq (r_2/r_1)^{-\lambda} M(r_1)$$

and

$$M(r_2)^{\log\{r_3/r_1\}} \leq (r_2/r_1)^{\log\{M(r_3)/M(r_1)\}} M(r_1)^{\log\{r_3/r_1\}}$$

and taking logarithms,

$$M(r_2)^{\log\{r_3/r_1\}} \leq M(r_1)^{\log\{r_3/r_2\}} M(r_3)^{\log\{r_2/r_1\}}. \quad \square$$

Clearly equality is achieved when $\phi(z)$ is constant, i.e., $f(z)$ is of the form Cz^u for some real u.

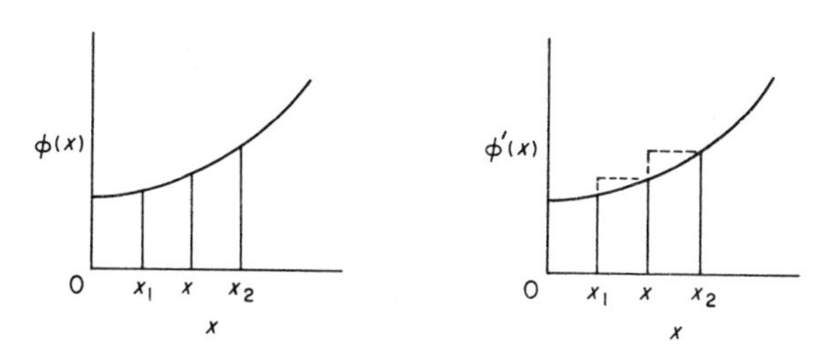

FIGURE 1

A sufficient condition for the convexity of $\phi(x)$ is that $\phi''(x) \geq 0$, i.e., $\phi'(x)$ is nondecreasing (Fig. 1). Thus for $x_1 < x < x_2$,

$$\frac{1}{x - x_1} \int_{x_1}^{x} \phi'(t)\, dt \leq \phi'(x) \leq \frac{1}{x_2 - x} \int_{x}^{x_2} \phi'(t)\, dt,$$

integration of which gives the previous definition of convexity. Alternatively, putting $x = \frac{1}{2}(x_1 + x_2)$ we obtain another more general definition of convexity, viz.,

$$\phi\{\tfrac{1}{2}x_1 + \tfrac{1}{2}x_2\} \leq \tfrac{1}{2}\{\phi(x_1) + \phi(x_2)\}.$$

Note that

$$(x - x_1)\phi'(x) \geq \int_{x_1}^{x} \phi'(t)\, dt$$

and

$$(x_2 - x)\phi'(x) \leq \int_{x}^{x_2} \phi'(t)\, dt.$$

Actually a lot more can be said about convex functions which are of particular interest in the theory of Harmonic functions. We state one or two results which are not difficult to prove.†

I. A function $f(x)$ continuous in an open interval I is convex if and only if for every pair of points $x_1, x_2 \in I$ and every pair of nonnegative numbers p_1, p_2, $p_1 + p_2 > 0$, we have

$$f\left(\frac{p_1 x_1 + p_2 x_2}{p_1 + p_2}\right) \leq \frac{p_1 f(x_1) + p_2 f(x_2)}{p_1 + p_2}.$$

This clearly reduces to our previous definition if $p_1 = p_2 = 1$.

† See S. Saks and A. Zygmund, "Analytic Functions," Monographie Matematyczne Vol. 28. Polska Akademia Nauk, Warsaw, 1965.

II. If $g(x)$ is continuous in a closed interval $[a, b]$ and $f(x)$ is convex in an open interval containing all the values of $g(x)$, then

$$f\left\{\frac{1}{b-a}\int_a^b g(x)\,dx\right\} \leq \frac{1}{b-a}\int_a^b f\{g(x)\}\,dx.$$

1.16 INFINITE PRODUCTS

In order to examine the behavior of entire functions, it is necessary to express them in as many different ways as possible. We will review properties of infinite products with a view to expressing entire functions in this form. Later we will study infinite series representation in some detail. Clearly some entire functions, the polynomials, have only finite product representation. A study of convergence and divergence of infinite products will now be undertaken.

Consider $(1 + a_1)(1 + a_2) \cdots$ containing an infinite number of factors, i.e., $\prod_{i=1}^{\infty} (1 + a_i)$, where no $a_i = -1$. Then let us write

$$P_n = \prod_{m=1}^{n} (1 + a_m).$$

1.16.1 Definition. The infinite product converges if there exists a finite L (not zero) such that

$$\lim_{n\to\infty} P_n = L.$$

If $P_n \to 0$, the product is said to *diverge to zero*. It is necessary that $a_n \to 0$ since $P_n = P_{n-1} + a_n P_{n-1}$, but clearly not sufficient since $\prod_{n=1}^{\infty}(1 + (1/n))$ diverges yet $a_n \to 0$.

1.16.2 Theorem. If $a_n \geq 0$ for all n, then $\prod_{n=1}^{\infty}(1 + a_n)$ and $\sum_{n=1}^{\infty} a_n$ converge or diverge together.

Proof. We note that P_n is a nondecreasing function of n (since each term is greater than or equal to one). Thus P_n either converges or approaches $+\infty$. We have

$$a_1 + a_2 + \cdots + a_n \leq (1 + a_1)(1 + a_2) \cdots (1 + a_n) \leq e^{a_1+a_2+\cdots+a_n}$$

and if P_n is bounded, then $\sum_{i=1}^{n} a_i$ is bounded. Also, if $\sum_{i=1}^{n} a_i$ is bounded, $\exp\{\sum_{i=1}^{n} a_i\}$ is bounded, and thus P_n is bounded. □

If $a_n \leq 0$ for all n, write $a_n = -b_n$ and consider $\prod_{n=1}^{\infty}(1 - b_n)$.

1.16.3 Theorem. If $b_n \geq 0$, $b_n \neq 1$ for all n, and $\sum_{n=1}^{\infty} b_n$ converges, then $\prod_{n=1}^{\infty}(1 - b_n)$ converges.

Proof. Since $\sum_{n=1}^{\infty} b_n$ converges, $\exists N$ such that

$$b_N + b_{N+1} + \cdots + b_n < \tfrac{1}{2}$$

and $b_n < 1$ for $n \geq N$. Thus

$$(1 - b_N)(1 - b_{N+1}) \geq 1 - b_N - b_{N+1}$$

and

$$\begin{aligned}(1 - b_N)(1 - b_{N+1})(1 - b_{N+2}) &\geq (1 - b_N - b_{N+1})(1 - b_{N+2}) \\ &\geq 1 - b_N - b_{N+1} - b_{N+2}.\end{aligned}$$

Therefore

$$(1 - b_N)(1 - b_{N+1}) \cdots (1 - b_n) \geq 1 - b_N \cdots - b_n > \tfrac{1}{2}$$

and P_n/P_{n-1} is decreasing [since for $n \gg N$, $(1 - b_n) < 1$] and has a positive lower limit. Therefore P_n/P_{n-1} tends to a positive limit, and since $P_{n-1} \neq 0$, P_n converges. $\square$

1.16.4 Theorem. If $0 \leq b_n < 1$ for all n and $\sum_{n=1}^{\infty} b_n$ diverges, then $\prod_{n=1}^{\infty}(1 - b_n)$ diverges to zero.

Proof. If $0 \leq b < 1$, $1 - b \leq e^{-b}$. Thus

$$(1 - b_1)(1 - b_2) \cdots (1 - b_n) \leq e^{-b_1 - b_2 - \cdots - b_n}$$

and the right-hand side approaches zero, hence the result. $\square$

1.16.5 Definition. The product $\prod_{n=1}^{\infty}(1 + a_n)$ is said to be *absolutely convergent* if $\prod_{n=1}^{\infty}(1 + |a_n|)$ converges. Thus a necessary and sufficient condition that the product should be absolutely convergent is that $\sum_{n=1}^{\infty} |a_n|$ should converge.

1.16.5.1 Lemma. An absolutely convergent product is convergent.

Proof. As before, write

$$P_n = \prod_{m=1}^{n} (1 + a_m)$$

and let

$$q_n = \prod_{m=1}^{n} (1 + |a_m|).$$

Then

$$P_n - P_{n-1} = (1 + a_1) \cdots (1 + a_{n-1})a_n$$

and

$$q_n - q_{n-1} = (1 + |a_1|) \cdots (1 + |a_{n-1}|) |a_n|.$$

Thus

$$|P_n - P_{n-1}| \leq q_n - q_{n-1}.$$

Now if $\prod_{n=1}^{\infty}(1 + |a_n|)$ converges, q_n approaches a limit and $\sum_{n=1}^{\infty}(q_n - q_{n-1})$ converges, and by the comparison theorem $\sum_{n=1}^{\infty}(P_n - P_{n-1})$ converges, i.e., P_n approaches a (nonzero) limit. ☐

We see that this limit cannot be zero, since if $\sum_{n=1}^{\infty} |a_n|$ converges and $1 + a_n \to 1$, the series $\sum_{n=1}^{\infty} |a_n/(1 + a_n)|$ converges. Thus the product

$$\prod_{m=1}^{\infty} \left(1 - \frac{a_m}{1 + a_m}\right)$$

approaches a limit, but this product is $1/P_n$, thus $\lim_{n\to\infty} P_n \neq 0$.

1.16.6 Theorem. A necessary and sufficient condition for the convergence of the infinite product $\prod_{n=1}^{\infty}(1 + a_n)$, a_n complex, is the convergence of $\sum_{n=1}^{\infty} \log(1 + a_n)$ where each log has its principal value.

Proof. Write

$$S_n = \sum_{r=1}^{n} \log(1 + a_r).$$

To establish sufficiency, we have that $P_n = \exp(S_n)$ and since the exponential function is continuous, $S_n \to S$ which implies $P_n \to e^S$. To establish necessity, we have that

$$S_n = \log P_n + 2q_n\pi i, \qquad q_n \text{ an integer.}$$

Since the principal value of the log of the product is not necessarily the sum of the principal values of its factors, q_n is not necessarily zero. We show that q_n is constant for all $n > N$ and from this necessity follows.

Write α_n, β_n the principal values of the arguments of $(1 + a_n)$ and P_n, respectively. If the infinite product converges, $\alpha_n \to 0$ (since $a_n \to 0$) and

$\beta_n \to \beta$, say, as $n \to \infty$. The integer q_n is then given by

$$\alpha_1 + \alpha_2 + \cdots + \alpha_n = \beta_n + 2\pi q_n$$

and

$$2(q_{n+1} - q_n)\pi = \alpha_{n+1} - (\beta_{n+1} - \beta_n) \to 0, \qquad n \to \infty.$$

However, since q_n is an integer, this implies $q_n = q$ for all sufficiently large n. Thus if P_n tends to the finite nonzero limit P as $n \to \infty$, it follows that

$$S_n \to \log P + 2q\pi i,$$

and the condition is necessary. □

1.16.7 Theorem. The product $\prod_{n=1}^{\infty}[1 + U_n(z)]$ is uniformly convergent in a region where the series $\sum_{n=1}^{\infty} | U_n(z) |$ converges uniformly.

Proof. Let M be an upper bound of the sum $\sum_{n=1}^{\infty} | U_n(z) |$ in the considered region. Then

$$\{1 + | U_1(z) |\}\{1 + | U_2(z) |\} \cdots \{1 + | U_n(z) |\} < e^{|U_1(z)+\cdots+U_n(z)|} \leq e^M.$$

Let

$$P_n(z) = \prod_{m=1}^{n} \{1 + | U_m(z) |\}.$$

Then

$$P_n(z) - P_{n-1}(z) = \{1 + | U_1(z) |\} \cdots \{1 + | U_{n-1}(z) |\} | U_n(z) | < e^M | U_n(z) |,$$

hence $\sum_{n=2}^{\infty}\{P_n(z) - P_{n-1}(z)\}$ converges uniformly since

$$\sum_{n=2}^{\infty} \{P_n(z) - P_{n-1}(z)\} < e^M \sum_{n=1}^{\infty} | U_n(z) |$$

and $\sum_{n=1}^{\infty} | U_n(z) |$ converges uniformly. Thus P_n approaches a limit and the absolutely convergent product converges. □

A considerable number of practical examples should be worked through in the standard texts, since there are many theorems or lemmas which could be cited, most of which are special cases of the previous theorems.†

† See in particular, the examples in E. C. Titchmarsh, "The Theory of Functions." Oxford Univ. Press, London and New York, 1939.

CHAPTER II

THE EXPANSION OF FUNCTIONS AND PICARD THEOREMS

As a preamble to the expansion of a meromorphic and an entire function, we recall the following.

2.1 RESIDUES

Let $f(z)$ have a pole of order k at $z = z_0$. Write

$$f(z) = \frac{a_{-k}}{(z - z_0)^k} + \cdots + \frac{a_{-1}}{z - z_0} + a_0 + \cdots$$

and

$$(z - z_0)^k f(z) = a_{-k} + \cdots + a_{-1}(z - z_0)^{k-1} + a_0(z - z_0)^k + \cdots.$$

Thus

$$\frac{d^{k-1}}{dz^{k-1}} [(z - z_0)^k f(z)] = (k - 1)!a_{-1} + \frac{k!}{1!} a_0(z - z_0)$$

$$+ \frac{(k + 1)!}{2!} a_1(z - z_0)^2 + \cdots,$$

$$\lim_{z \to z_0} \left\{ \frac{d^{k-1}}{dz^{k-1}} [(z - z_0)^k f(z)] \right\} = (k - 1)!a_{-1},$$

and

$$a_{-1} = \text{res}(z_0) = \frac{1}{(k-1)!} \lim_{z \to z_0} \left\{ \frac{d^{k-1}}{dz^{k-1}} [(z - z_0)^k f(z)] \right\}.$$

2.2 EXPANSION OF A MEROMORPHIC FUNCTION

Since a meromorphic function is analytic in a region except possibly at a finite number of poles, we have the following theorem.

2.2.1 Theorem. Let $f(z)$ be a function whose only singularities, except at infinity, are poles. Suppose all poles are simple. Let them be $a_1, a_2, \ldots$ and be ordered such that $0 < |a_1| \leq |a_2| \leq \cdots$ and let the residues at the poles be $b_1, b_2, \ldots$. Suppose there is a sequence of contours C_n, such that C_n includes $a_1, a_2, \ldots, a_n$, but no other poles. Let the minimum distance R_n of C_n from the origin approach ∞ with n, while L_n the length of C_n is $O(R_n)$ and such that on C_n, $f(z) = o(R_n)$. This last condition is satisfied if, for example, $f(z)$ is bounded on all C_n. Then

$$f(z) = f(0) + \sum_{n=1}^{\infty} b_n \left(\frac{1}{z - a_n} + \frac{1}{a_n} \right)$$

for all z except poles.

Proof. Consider

$$I = \frac{1}{2\pi i} \int_{C_n} \frac{f(w)}{w(w-z)}\, dw$$

where $z \in C_n$. The integrand has poles at a_m, 0, and $w = z$, with residues $b_m/a_m(a_m - z)$, $-f(0)/z, f(z)/z$, respectively

$$I = 2\pi i \sum \text{residues} = \sum_{m=1}^{n} \frac{b_m}{a_m(a_m - z)} - \frac{f(0)}{z} + \frac{f(z)}{z}$$

and

$$|I| \leq \frac{1}{2\pi} \frac{L_n}{R_n(R_n - |z|)} \cdot \max_{C_n} |f(w)| \to 0 \qquad \text{as} \quad n \to \infty,$$

by the conditions imposed. Thus

$$\begin{aligned} f(z) &= f(0) - \lim_{n \to \infty} \sum_{m=1}^{n} \frac{z b_m}{a_m(a_m - z)} \\ &= f(0) + \sum_{i=1}^{\infty} b_i \left(\frac{1}{z - a_i} + \frac{1}{a_i} \right). \end{aligned}$$

Note, that the series converges uniformly inside any closed contour such that all poles are inside. □

2.3 POLES AND ZEROS OF A MEROMORPHIC FUNCTION

2.3.1 Theorem. If $f(z)$ is analytic in and on a closed contour C apart from a finite number of poles and if $f(z) \neq 0$, $z \in C$, then

$$\frac{1}{2\pi i} \int_C \frac{f'(z)}{f(z)} \, dz = N - P$$

where N is the number of zeros in C (zero of order m counted m times) and P is the number of poles in C (pole of order m counted m times).

Proof. Let $z = a$ be a zero of order m. Then in the neighborhood of $z = a$,

$$f(z) = (z - a)^m g(z) \qquad \text{for} \quad g(z) \neq 0$$

and analytic in the neighborhood of $z = a$. Thus

$$\frac{f'(z)}{f(z)} = \frac{m}{z - a} + \frac{g'(z)}{g(z)}.$$

The last term is analytic at $z = a$, therefore $f'(z)/f(z)$ has a simple pole at $z = a$, with residue m.

The sum of the residues at the zeros of $f(z)$ is N (the number of zeros). Similarly by writing $-m$ for m, the sum of the residues at the poles of $f(z)$ is $-P$ (number of poles). If $f(z)$ is an entire function, the number of zeros

$$N = \frac{1}{2\pi i} \int_C \frac{f'(z)}{f(z)} \, dz. \quad \square$$

2.3.2. Corollary. If $\phi(z)$ is analytic in and on C and if $f(z)$ has zeros at $a_1, \ldots, a_p$ and poles at $b_1, \ldots, b_q$, multiplicity being counted as before,

$$\frac{1}{2\pi i} \int_C \frac{f'(z)}{f(z)} \phi(z) \, dz = \sum_{\mu=1}^{p} \phi(a_\mu) - \sum_{\nu=1}^{q} \phi(b_\nu),$$

for if $z = a$ is a zero of order m, we have in the neighborhood of $z = a$, $f(z) = (z - a)^m g(z)$ where $g(z) \neq 0$ and analytic. Thus

$$\frac{f'(z)}{f(z)} \phi(z) = \frac{m\phi(z)}{z - a} + \frac{g'(z)}{g(z)} \phi(z).$$

The last term is analytic at $z = a$, therefore the left-hand side has a simple pole at $z = a$ with residue $m\phi(a)$. Applying the previous theorem we have the result.

2.4 EXPANSION OF AN ENTIRE FUNCTION AS AN INFINITE PRODUCT

Suppose $f(z)$ has simple zeros at the points $a_1, a_2, \ldots, a_n$. In the neighborhood of a_n, $f(z) = (z - a_n)g(z)$, where $g(z)$ is analytic and nonzero. Thus

$$\frac{f'(z)}{f(z)} = \frac{1}{z - a_n} + \frac{g'(z)}{g(z)}$$

with $g'(z)/g(z)$ being analytic at a_n. Hence $f'(z)/f(z)$ has a simple pole at $z = a_n$ with residue 1.

Suppose $f'(z)/f(z)$ is a function of the type considered in the expansion of a meromorphic function $f(z)$. Then $P(z) = f'(z)/f(z)$ has poles at a_1, $a_2, \ldots, a_n$ and

$$P(z) = P(0) + \sum_{n=1}^{\infty} b_n\left(\frac{1}{z - a_n} + \frac{1}{a_n}\right).$$

Clearly $P(0) = f'(0)/f(0)$ and $b_n = 1$ for all n. Thus

$$\frac{f'(z)}{f(z)} = \frac{f'(0)}{f(0)} + \sum_{n=1}^{\infty}\left(\frac{1}{z - a_n} + \frac{1}{a_n}\right).$$

Integrating from 0 to z along a path not passing through a pole, we have

$$\log f(z) - \log f(0) = z\,\frac{f'(0)}{f(0)} + \sum_{n=1}^{\infty}\left\{\log(z - a_n) - \log(-a_n) + \frac{z}{a_n}\right\}$$

where the value of the logarithms depend upon the path. Taking exponentials,

$$f(z) = f(0)\exp\left(z\,\frac{f'(0)}{f(0)}\right)\prod_{n=1}^{\infty}\left(1 - \frac{z}{a_n}\right)\exp\frac{z}{a_n}.$$

As an example, consider

$$f(z) = \frac{\sin z}{z} = \prod_{n=-\infty}^{\infty}{}' \left(1 - \frac{z}{n\pi}\right)\exp\frac{z}{n\pi},$$

or

$$\sin z = z\prod_{n=1}^{\infty}\left(1 - \frac{z^2}{n^2\pi^2}\right).$$

The next theorem called *Rouché's theorem* follows somewhat naturally.

2.5 Theorem. If $f(z)$ and $g(z)$ are analytic in and on a closed contour C, and if $|g(z)| < |f(z)|$ on C, then $f(z)$ and $f(z) + g(z)$ have the same number of zeros inside C.

Proof. Let $\phi(z) = g(z)/f(z)$. Then $|\phi(z)| < 1$ on C. Note that neither $f(z)$ nor $f(z) + g(z)$ has a zero on C. If the number of zeros of $f(z) + g(z)$ inside C is denoted by N', then

$$\begin{aligned}
N' &= \frac{1}{2\pi i}\int_C \frac{f' + g'}{f + g}\,dz \\
&= \frac{1}{2\pi i}\int_C \frac{f' + f'\phi + f\phi'}{(1 + \phi)f}\,dz \\
&= \frac{1}{2\pi i}\int_C \frac{f'}{f}\,dz + \frac{1}{2\pi i}\int_C \frac{\phi'}{1 + \phi}\,dz \\
&= N + \frac{1}{2\pi i}\int_C \phi'\{1 - \phi + \phi^2 - \cdots\}\,dz.
\end{aligned}$$

Also,

$$\int_C \phi'\phi^n\,dz = \frac{\phi^{n+1}}{n+1}\Big]_C = 0,$$

thus by uniform convergence of the series we have $N' = N$. □

The following theorem is now a consequence.

2.6 A Theorem of Hurwitz. Let $f_n(z)$ be a sequence of functions analytic in a region D bounded by a simple closed contour. Let $f_n(z)$ tend to $f(z)$ uniformly. Assume $f(z) \not\equiv 0$, and let z_0 be an interior point of D. Then z_0 is a zero of $f(z)$ if and only if z_0 is a limit point of zeros of $f_n(z)$, $n = 1, 2, \ldots$ and points which are zeros for infinite n are counted as limit points.

Proof. Choose ϱ small and such that the circle $|z - z_0| = \varrho$ is in D and contains or has on it no zero of $f(z)$ except possibly z_0. Then $|f(z)|$ has a positive lower bound on the circle, i.e.,

$$|f(z)| \geq m > 0.$$

Having fixed ϱ and m, choose N so large that

$$|f_n(z) - f(z)| < m \qquad \text{for} \quad n \geq N, \quad z \in \{|z - z_0| = \varrho\}.$$

Since

$$f_n(z) = f(z) + \{f_n(z) - f(z)\},$$

$f_n(z)$ has the same number of zeros in the circle as $f(z)$. This follows from Rouchés theorem because $f(z)$ and $f_n(z) - f(z)$ are analytic in and on a closed contour C in a region D and $|f_n(z) - f(z)| < |f(z)|$. Thus if $f(z_0) = 0$, then $f_n(z)$ has exactly one zero in C for all $n \geq N$ so that z_0 is a limit point of zeros of $f_n(z)$. Also, if $f(z_0) \neq 0$, then $f_n(z_0) \neq 0$ in C. □

2.7 The next group of theorems although elementary in nature, examine more carefully the relation between order and the number of roots of an equation of the form $f(z) = AP(z)$ where $f(z)$ is entire, A is a constant, and $P(z)$ is a nonzero polynomial.

The only deficiency in the theorems is that they do not deal with functions of infinite order, a ramification which will be dealt with later on in the survey.

2.7.1 Picard Theorems. It will be useful to establish first of all, a few lemmas concerning inequalities for coefficients of power series.

Consider the entire function $f(z) = a_0 + a_1 z + \cdots + a_n z^n + \cdots$. Write

$$a_n = \alpha_n + i\beta_n$$

and

$$z = r(\cos\theta + i\sin\theta), \qquad r \geq 0,$$

Then

$$f(z) = \sum_{n=0}^{\infty} (\alpha_n + i\beta_n) r^n (\cos\theta + i\sin\theta)^n.$$

If we write

$$f(z) = U(r, \theta) + iV(r, \theta),$$

then

$$U(r, \theta) = \alpha_0 + \sum_{n=1}^{\infty} (\alpha_n \cos n\theta - \beta_n \sin n\theta) r^n. \tag{1}$$

Fixing r and integrating with respect to θ from 0 to 2π, we have

$$\alpha_0 = \frac{1}{2\pi} \int_0^{2\pi} U(r, \theta)\, d\theta.$$

To compute α_p, $p \geq 1$ multiply (1) by $\cos p\theta$ and integrate with respect to θ from 0 to 2π. Thus

$$\int_0^{2\pi} U(r, \theta) \cos p\theta\, d\theta = a_p r^p \int_0^{2\pi} \cos^2 p\theta\, d\theta,$$

from which

$$\alpha_p = \frac{1}{\pi r^p} \int_0^{2\pi} U(r, \theta) \cos p\theta \, d\theta \qquad p = 1, 2, \ldots$$

and similarly

$$\beta_p = \frac{1}{\pi r^p} \int_0^{2\pi} U(r, \theta) \sin p\theta \, d\theta, \qquad p = 1, 2, \ldots .$$

Also,

$$2\alpha_0 \pm \alpha_p r^p = \frac{1}{\pi} \int_0^{2\pi} U(r, \theta)(1 \pm \cos p\theta) \, d\theta$$

and

$$2\alpha_0 \pm \beta_p r^p = \frac{1}{\pi} \int_0^{2\pi} U(r, \theta)(1 \pm \sin p\theta) \, d\theta .$$

Note that the factors multiplying $U(r, \theta)$ are now nonnegative.

Let $\mu(r) = \max_{0 \leq \theta \leq 2\pi} U(r, \theta)$ on the circle of radius r. Then

$$2\alpha_0 \pm \alpha_p r^p \leq \frac{\mu(r)}{\pi} \int_0^{2\pi} (1 \pm \cos p\theta) \, d\theta = 2\mu(r).$$

Similarly

$$2\alpha_0 \pm \beta_p r^p \leq 2\mu(r).$$

Hence

$$|\alpha_p| \leq \frac{2\{\mu(r) - \alpha_0\}}{r^p} \qquad \text{and} \qquad |\beta_p| \leq \frac{2\{\mu(r) - \alpha_0\}}{r^p}.$$

2.7.2 Lemma. If $U(r, \theta)$ the real part of an entire function $f(z)$ satisfies

$$U(r, \theta) \leq \mu(r) \leq Cr^\delta, \qquad \delta > 0,$$

for all $r > N$, then $f(z)$ is a polynomial of degree not exceeding $n = [\delta]$, i.e., the integer part of δ.

Proof. Since $|\alpha_p|$ and $|\beta_p| \leq 2\{\mu(r) - \alpha_0\}/r^p$, we have that

$$|\alpha_p| \leq \frac{2(Cr^\delta - \alpha_0)}{r^p} \qquad \text{and} \qquad |\beta_p| \leq \frac{2(Cr^\delta - \alpha_0)}{r^p}.$$

If $p > [\delta]$, since p is an integer, $p \geq [\delta] + 1 > \delta$ hence $\alpha_p, \beta_p \to 0$, $r \to \infty$, and $\alpha_p + i\beta_p = a_p = 0$ for $p > [\delta] = n$. □

2.7.3 Lemma. If $f(z)$ is a transcendental entire function and $P(z)$ and $Q(z)$ are polynomials of degree m, n, respectively and if $P(z) \not\equiv 0$, then the order ϱ_1 of $P(z)f(z) + Q(z)$ coincides with the order ϱ of $f(z)$, i.e.,

$$\varrho_1 = \varrho .$$

Proof. Writing

$$M(r) = \max_{|z| \leq r} |f(z)| \qquad \text{and} \qquad M_1(r) = \max_{|z| \leq r} |P(z)f(z) + Q(z)|,$$

we confine ourselves to values of the functions at points on the circle $|z| = r$.

Let $a_m z^m$, $b_n z^n$ denote, respectively, the leading terms in the polynomials $P(z)$ and $Q(z)$. Taking $\varepsilon = \frac{1}{2}$ in the inequalities for polynomials previously evaluated, we assert for $|z| = r > r_0$, that

$$\tfrac{1}{2} |a_m| r^m \leq |P(z)| \leq \tfrac{3}{2} |a_m| r^m$$

and

$$\tfrac{1}{2} |b_n| r^n \leq |Q(z)| \leq \tfrac{3}{2} |b_n| r^n.$$

Let z_0 be a point on $|z| = r$ at which $|f(z)|$ attains its maximum $M(r)$. Then

$$\begin{aligned} M_1(r) &\geq |P(z_0)f(z_0) + Q(z_0)| \geq |P(z_0)|\,|f(z_0)| - |Q(z_0)| \\ &\geq \tfrac{1}{2} |a_m| r^m M(r) - \tfrac{3}{2} |b_n| r^n. \end{aligned}$$

However, at z_1 (on the same circle) at which $|P(z)f(z) + Q(z)|$ attains its maximum $M_1(r)$, we have

$$M_1(r) = |P(z_1)f(z_1) + Q(z_1)| \leq \tfrac{3}{2} |a_m| r^m M(r) + \tfrac{3}{2} |b_n| r^n.$$

Thus

$$\begin{aligned} \frac{1}{2} |a_m| r^m M(r) \left\{ 1 - \frac{3 |b_n| r^n}{|a_m| r^m M(r)} \right\} &\leq M_1(r) \\ &\leq \frac{3}{2} |a_m| r^m M(r) \left\{ 1 + \frac{|b_n| r^n}{|a_m| r^m M(r)} \right\}. \end{aligned}$$

Since $f(z)$ is a transcendental entire function it has been proved that $M(r)$ increases faster than the maximum modulus of any polynomial and hence faster than any power of r. Therefore each $\{\cdots\} \to 1$ as $r \to \infty$ and hence for r sufficiently large, the right-hand $\{\cdots\} < 2$ and the left hand $\{\cdots\} > \frac{2}{3}$, i.e.,

$$\tfrac{1}{3} |a_m| r^m M(r) \leq M_1(r) \leq 3 |a_m| r^m M(r).$$

We now require

$$\overline{\lim_{r\to\infty}} \frac{\log \log M(r)}{\log r},$$

$$\log(\tfrac{1}{3} \mid a_m \mid r^m) + \log M(r) \leq \log M(r_1) \leq \log(3 \mid a_m \mid r^m) + \log M(r),$$

i.e.,

$$\log M(r)\left\{1 + \frac{\log(\frac{1}{3} \mid a_m \mid r^m)}{\log M(r)}\right\} \leq \log M(r_1) \leq \log M(r)\left\{1 + \frac{\log(3 \mid a_m \mid r^m)}{\log M(r)}\right\}$$

and each $\{\cdots\} \to 1$ as $r \to \infty$. Hence the right-hand $\{\cdots\} < 2$ and the left-hand $\{\cdots\} > \frac{1}{2}$, and

$$\tfrac{1}{2} \log M(r) \leq \log M_1(r) \leq 2 \log M(r).$$

Taking logarithms again,

$$\overline{\lim_{r\to\infty}} \frac{\log \log M(r)}{\log r} \leq \overline{\lim_{r\to\infty}} \frac{\log \log M_1(r)}{\log r} \leq \overline{\lim_{r\to\infty}} \frac{\log \log M(r)}{\log r},$$

consequently $\varrho = \varrho_1$ and the orders of $f(z)$ and $P(z)f(z) + Q(z)$ are equal $[P(z) \not\equiv 0]$. □

2.7.4 Lemma. The order of an entire function

$$f(z) = P(z)\, e^{g(z)} + Q(z),$$

where $P(z)$, $Q(z)$, and $g(z)$ are polynomials and $P(z) \not\equiv 0$, is equal to the degree of $g(z)$.

Proof. From the previous lemma, the order ϱ of $f(z)$ coincides with the order of $\phi(z) = e^{g(z)}$. We need to show that

$$\overline{\lim_{r\to\infty}} \frac{\log \log M_1(r)}{\log r} = n$$

where $\max_{|z|=r} \mid \phi(z) \mid = M_1(r)$ and n is the degree of $g(z)$. Set

$$g(z) = c_0 + c_1 z + \cdots + c_n z^n,$$
$$c_k = \varrho_k(\cos \alpha_k + i \sin \alpha_k)$$

and

$$z = r(\cos \theta + i \sin \theta).$$

By hypothesis, $\varrho_n = |c_n| \neq 0$, thus

$$g(z) = \sum_{k=0}^{n} \varrho_k r^k (\cos \alpha_k + i \sin \alpha_k)(\cos k\theta + i \sin k\theta)$$
$$= \sum_{k=0}^{n} \varrho_k r^k \{\cos(\alpha_k + k\theta) + i \sin(\alpha_k + k\theta)\}$$

and

$$|\phi(z)| = \exp \sum_{k=0}^{n} \varrho_k r^k \cos(\alpha_k + k\theta)$$

and

$$\log |\phi(z)| = \sum_{k=0}^{n} \varrho_k r^k \cos(\alpha_k + k\theta).$$

We require now $\log \max_{|z|=r} |\phi(z)|$. For fixed r and $0 \leq \theta \leq 2\pi$

$$\log |\phi(z)| \leq \sum_{k=0}^{n} \varrho_k r^k = \varrho_n r^n \left(1 + \frac{\varrho_{n-1}}{\varrho_n} \frac{1}{r} + \cdots + \frac{\varrho_0}{\varrho_n} \frac{1}{r^n}\right),$$

hence for ε in $(0, 1)$ and $r > r(\varepsilon)$,

$$\log |\phi(z)| < \varrho_n r^n (1 + \varepsilon)$$

and

$$\log M_1(r) < \varrho_n r^n (1 + \varepsilon).$$

Let z_0 be a point on $|z| = r$ such that $\cos(\alpha_n + n\theta) = 1$ (actually there are n such points). Then

$$\log M_1(r) \geq \log |\phi(z_0)| = \varrho_n r^n + \sum_{k=0}^{n-1} \varrho_k r^k \cos(\alpha_k + k\theta)$$
$$\geq \varrho_n r^n - \sum_{k=0}^{n-1} \varrho_k r^k$$
$$= \varrho_n r^n \left\{1 - \left(\frac{\varrho_{n-1}}{\varrho_n} \frac{1}{r} + \cdots + \frac{\varrho_0}{\varrho_n} \frac{1}{r^n}\right)\right\}$$
$$> \varrho_n r^n (1 - \varepsilon) \qquad \text{for} \quad r > r(\varepsilon).$$

Thus

$$\varrho_n r^n (1 - \varepsilon) < \log M_1(r) < \varrho_n r^n (1 + \varepsilon)$$

and

$$\overline{\lim_{r \to \infty}} \frac{\log \log M_1(r)}{\log r} = n. \quad \square$$

2.7.5 Lemma. If $g(z)$ is an entire function and if the order of the function $f(z) = e^{g(z)}$ is finite, then $g(z)$ is a polynomial and hence the order of $f(z)$ is an integer.

Proof. If $z = r(\cos\theta + i\sin\theta)$ and $U(r, \theta)$ is the real part of $g(z)$, then

$$\log |f(z)| = U(r, \theta).$$

Let

$$\max_{|z|=r} |f(z)| = M(r) \qquad \text{and} \qquad \max_{0\leq\theta\leq 2\pi} U(r, \theta) = \mu(r).$$

Then

$$\log M(r) = \mu(r). \tag{1}$$

Suppose δ is the order of $f(z)$. Then

$$\overline{\lim_{r\to\infty}} \frac{\log\log M(r)}{\log r} = \delta.$$

Given $\varepsilon > 0$, $\exists r(\varepsilon) > 1$ such that for $r > r(\varepsilon)$

$$\frac{\log\log M(r)}{\log r} < \delta + \varepsilon$$

i.e.,

$$\log\log M(r) < \log(r^{\delta+\varepsilon})$$

or

$$\log M(r) < r^{\delta+\varepsilon}.$$

Thus from (1),

$$\mu(r) < r^{\delta+\varepsilon} \qquad \text{for} \quad r > r(\varepsilon).$$

From Lemma 2.7.2 it follows that $g(z)$ is a polynomial the degree n of which satisfies $n \leq [\delta + \varepsilon]$ and ε is arbitrarily small. Thus $n \leq [\delta]$ and (by Lemma 2.7.4) the order of $f(z)$ coincides with n, i.e., δ is integral. □

2.7.6 Theorem. Let $f(z)$ denote a transcendental entire function, the order δ of which is finite but nonintegral. Then if $P(z)$ is a nonzero polynomial, the equation

$$f(z) = AP(z)$$

has infinitely many roots for every complex number A (no exceptions).

Proof. Suppose $\exists A = A_0$ for which $f(z) = A_0 P(z)$ has only finitely many roots if any at all. Then the entire function $f(z) - A_0P(z)$ has only finitely many zeros. Therefore by a previous theorem we may write

$$f(z) - A_0P(z) = Q(z)e^{g(z)}$$

where $Q(z)$ is a nonzero polynomial (equal to 1 if $f - A_0P$ has no zeros). Further $g(z)$ is an entire function, and we have

$$f(z) = A_0P(z) + Q(z)e^{g(z)}.$$

Thus the order δ of $f(z)$ coincides with the order of $e^{g(z)}$ which, being finite by a previous lemma, is integral. The contradiction proves the theorem. □

2.7.7 Theorem. If $f(z)$ is a transcendental entire function of finite integral order n and if $P(z)$ is a nonzero polynomial, then

$$f(z) = AP(z)$$

has infinitely many roots for every complex number A, with the possible exception of one value.

Proof. Suppose there exists at least two values a, b at which $f(z) = AP(z)$ has only finitely many roots. Then $f(z) - aP(z)$ and $f(z) - bP(z)$ have only finitely many zeros. Thus

$$f(z) - aP(z) = Q_1(z)e^{g_1(z)} \tag{1}$$

and

$$f(z) - bP(z) = Q_2(z)e^{g_2(z)} \tag{2}$$

where $Q_1(z)$ and $Q_2(z)$ are nonzero polynomials and $g_1(z)$ and $g_2(z)$ are entire functions. Hence the orders of $e^{g_1(z)}$ and $e^{g_2(z)}$ coincide with the order n of $f(z)$. We also conclude that $g_1(z)$ and $g_2(z)$ are polynomials of degree n. Hence $n \geq 1$ since for $n = 0$, $g_1(z)$ and $g_2(z)$ would be constants and $f(z)$ would be a polynomial but not a transcendental entire function. Subtracting (1) from (2),

$$Q_1(z)e^{g_1(z)} - Q_2(z)e^{g_2(z)} = (b - a)P(z) = R(z),$$

where $R(z)$ is a nonzero polynomial since $b \neq a$ and by hypothesis $P(z) \neq 0$. We need to show that this equation is not possible for $Q_1(z)$, $Q_2(z)$ and $R(z)$ nonzero polynomials and $g_1(z)$, $g_2(z)$ polynomials of degree greater than or equal to 1. Differentiating

$$(Q_1' + Q_1g_1')e^{g_1} - (Q_2' + Q_2g_2')e^{g_2} = R'$$

also

$$Q_1 e^{g_1} - Q_2 e^{g_2} = R.$$

Regarding these as a system of two equations in the unknowns e^{g_1} and e^{g_2}, the determinant of the system $\Delta(z)$ is given by

$$\begin{aligned}\Delta(z) &= Q_1(Q_2' + Q_2 g_2') - Q_2(Q_1' + Q_1 g_1') \\ &= -Q_2 Q_1' + Q_1 Q_2' - Q_1 Q_2(g_1' - g_2').\end{aligned}$$

We must now show that $\Delta(z) \not\equiv 0$. If we assume that $\Delta(z) = 0$, divide by $-Q_1 Q_2$ to obtain

$$\frac{Q_1'}{Q_1} - \frac{Q_2'}{Q_2} + (g_1 - g_2)' = 0.$$

Integrating this equation we have

$$\log \frac{Q_1}{Q_2} + g_1 - g_2 = \text{constant} = C_1.$$

Thus

$$\frac{Q_1}{Q_2} e^{g_1 - g_2} = e^{C_1} = C \neq 0.$$

However, by dividing Eqs. (1) and (2) we obtain

$$\frac{Q_1}{Q_2} e^{g_1 - g_2} = \frac{f - aP}{f - bP}.$$

Thus

$$\frac{f - aP}{f - bP} = C$$

and

$$f(z)(1 - C) = P(z)(a - bC).$$

Since $b \neq a$, we have $C \neq 1$ and hence

$$f(z) = \frac{a - bC}{1 - C} P(z)$$

which contradicts the hypothesis, for $f(z)$ is a transcendental entire function and $P(z)$ is a polynomial. Thus $\Delta(z) \not\equiv 0$. Solving for e^{g_1} and e^{g_2}, we obtain

$$e^{g_1} = \frac{-R(Q_2' + Q_2 g_2') + R' Q_2}{\Delta(z)} \quad \text{and} \quad e^{g_2} = \frac{-R(Q_1' + Q_1 g_1') + R' Q_1}{\Delta(z)}.$$

However, these equations contain a contradiction since the left-hand side is a transcendental entire function and the function on the right-hand side is a rational function, and thus a polynomial since it is entire. We can say [for $P(z) = 1$] that $f(z)$ takes on all values an infinite number of times with one possible exception. □

The previous two theorems are special cases of the following theorem.

2.7.8 Theorem. Let $Q(z)$ be a transcendental meromorphic function. For every complex number A (finite or infinite) with two possible exceptions, $Q(z) = A$ has infinitely many roots.

Proof. If $Q(z)$ is an entire function, it does not become infinite at any z. Thus $A = \infty$ is an exceptional value for every entire function. Consequently, on the basis of the previous theorem there may exist at most *one* finite exceptional value. If $f(z)$ is a transcendental entire function and $P(z)$ is a nonzero polynomial, the meromorphic function $Q(z) = f(z)/P(z)$ becomes infinite only at a finite number of points, namely the zeros of $P(z)$. Thus ∞ is an exceptional value, and further, one other finite exceptional value is possible. Thus $f(z)/P(z) = A$ or $f(z) = AP(z)$ has infinitely many roots for every A except possibly a single finite value of A. Let $f(z)$ and $g(z)$ denote two transcendental entire functions whose ratio is not rational. Hence $f(z)/g(z)$ is meromorphic. Thus $f(z)/g(z) = A$ has infinitely many roots with two possible exceptions. □

CHAPTER III

THEOREMS CONCERNING THE MODULUS OF A FUNCTION AND ITS ZEROS

It will now be expedient to study the real part of an analytic function. Several theorems transpire, the most important of which is Jensen's theorem which relates the modulus of a function to its zeros. The Poisson–Jensen theorem is somewhat of a starting point for Nevanlinna theory which is in effect a much more delicate analysis of the behavior of meromorphic functions.

We examine a theorem similar to Cauchy's inequality but involving the upper bound of $\mathscr{R}\{f(z)\}$ on $|z| = r$ rather than $M(r)$. Let $A(r)$ be the upper bound of the real part of $f(z)$ on $|z| = r$.

3.1 Theorem. If $f(z) = \sum_{n=0}^{\infty} a_n z^n$, then

$$|a_n| r^n \leq \max\{4A(r), 0\} - 2\mathscr{R}\{f(0)\}$$

for all $n > 0$ and r.

Proof. Let $z = re^{i\theta}$ and

$$f(z) = \sum_{n=0}^{\infty} a_n z^n = U(r, \theta) + iV(r, \theta)$$

and let $a_n = \alpha_n + i\beta_n$. Then

$$U(r, \theta) = \sum_{n=0}^{\infty} (\alpha_n \cos n\theta - \beta_n \sin n\theta) r^n.$$

The series converges uniformly with respect to θ. Thus we may multiply by $\cos n\theta$ or $\sin n\theta$ and integrate term by term. Thus

$$\int_0^{2\pi} U(r, \theta) \cos n\theta \, d\theta = \int_0^{2\pi} \alpha_n r^n \cos^2 n\theta \, d\theta = \pi \alpha_n r^n.$$

Similarly

$$\int_0^{2\pi} U(r, \theta) \sin n\theta \, d\theta = -\pi \beta_n r^n, \qquad n > 0;$$

also,

$$\int_0^{2\pi} U(r, \theta) \, d\theta = 2\pi \alpha_0.$$

Hence

$$a_n r^n = (\alpha_n + i\beta_n) r^n = \frac{1}{\pi} \int_0^{2\pi} U(r, \theta) e^{-in\theta} \, d\theta, \qquad n > 0$$

and

$$| a_n | r^n \leq \frac{1}{\pi} \int_0^{2\pi} | U(r, \theta) | \, d\theta.$$

Thus

$$| a_n | r^n + 2\alpha_0 \leq \frac{1}{\pi} \int_0^{2\pi} \{ | U(r, \theta) | + U(r, \theta) \} \, d\theta. \tag{1}$$

Note, $| u | + u = 0$ for $u < 0$. Hence if $A(r) < 0$, the right-hand side of (1) becomes zero and if $A(r) \geq 0$, the right-hand side does not exceed

$$\frac{1}{\pi} \int_0^{2\pi} 2A(r) \, d\theta = 4A(r).$$

Thus

$$| a_n | r^n \leq \max\{4A(r), 0\} - 2\mathscr{R}\{f(0)\}. \quad \square$$

This may be improved as follows. Write

$$a_n = \alpha_n + i\beta_n = | a_n | e^{i\theta_n},$$

then

$$\begin{aligned}|a_n| r^n &= (\alpha_n + i\beta_n) r^n e^{-i\theta_n}\\ &= \frac{1}{\pi}\int_0^{2\pi} U(r, \theta) e^{-in\theta} e^{-i\theta_n}\, d\theta\\ &= \frac{1}{\pi}\int_0^{2\pi} U(r, \theta)\cos(n\theta + \theta_n)\, d\theta \qquad (|a_n| r^n \text{ real}).\end{aligned}$$

Thus

$$\begin{aligned}|a_n| r^n + 2\alpha_0 &= \frac{1}{\pi}\int_0^{2\pi} U(r, \theta)\{1 + \cos(n\theta + \theta_n)\}\, d\theta\\ &\leq \frac{1}{\pi}\int_0^{2\pi} A(r)\{1 + \cos(n\theta + \theta_n)\}\, d\theta\\ &= \frac{A(r)}{\pi}\int_0^{2\pi} \{1 + \cos(n\theta + \theta_n)\}\, d\theta\\ &= 2A(r)\end{aligned}$$

and

$$|a_n| r^n \leq 2[A(r) - \mathscr{R}\{f(0)\}].$$

3.2 POISSON'S INTEGRAL FORMULA

3.2.1 Theorem. Let $f(z)$ be analytic in a region D including $|z| \leq R$ and let $U(r, \theta)$ be $\mathscr{R}\{f(z)\}$. Then for $0 \leq r < R$

$$U(r, \theta) = \frac{1}{2\pi}\int_0^{2\pi} \frac{R^2 - r^2}{R^2 - 2Rr\cos(\theta - \phi) + r^2} U(R, \phi)\, d\phi.$$

[Similarly for $V(r, \theta) = \mathscr{I}\{f(z)\}$.]

Proof. Let

$$f(z) = \sum_{n=0}^{\infty} (\alpha_n + i\beta_n) r^n e^{in\theta}, \qquad r \leq R.$$

We have

$$U(r, \theta) = \sum_{n=0}^{\infty} (\alpha_n \cos n\theta - \beta_n \sin n\theta) r^n,$$

$$\alpha_n R^n = \frac{1}{\pi}\int_0^{2\pi} U(R, \phi)\cos n\phi\, d\phi, \qquad n > 0,$$

$$\beta_n R^n = -\frac{1}{\pi}\int_0^{2\pi} U(R, \phi)\sin n\phi\, d\phi, \qquad n > 0,$$

and

$$\alpha_0 = \frac{1}{2\pi} \int_0^{2\pi} U(R, \phi)\, d\phi.$$

$$U(r, \theta) = \frac{1}{2\pi} \int_0^{2\pi} U(R, \phi)\, d\phi$$

$$+ \frac{1}{\pi} \sum_{n=1}^{\infty} \frac{r^n}{R^n} \int_0^{2\pi} U(R, \phi)\{\cos n\theta \cos n\phi + \sin n\theta \sin n\phi\}\, d\phi$$

$$= \frac{1}{\pi} \int_0^{2\pi} U(R, \phi)\left\{\frac{1}{2} + \sum_{n=1}^{\infty} \cos n(\theta - \phi)\left(\frac{r}{R}\right)^n\right\} d\phi. \quad \square$$

Inversion of the summation and integral process is justified by uniform convergence. To sum

$$\sum_{n=1}^{\infty} \cos n(\theta - \phi)(r/R)^n,$$

use

$$C = \sum_{n=1}^{\infty} \cos n\alpha \cdot x^n \qquad \text{and} \qquad iS = i \sum_{n=1}^{\infty} \sin n\alpha \cdot x^n.$$

Thus

$$C = \frac{x \cos \alpha - x^2}{1 - 2x \cos \alpha + x^2}$$

and adding $\frac{1}{2}$ we obtain Poisson's formula.

3.3 Jensen's Theorem. Let $f(z)$ be analytic for $|z| < R$. Suppose $f(0) \neq 0$ and let $r_1, r_2, \ldots, r_n, \ldots$ be the moduli of the zeros of $f(z)$ in $|z| < R$ arranged as a nondecreasing sequence. If $r_n \leq r \leq r_{n+1}$, then

$$\log \frac{r^n |f(0)|}{r_1 r_2 \cdots r_n} = \frac{1}{2\pi} \int_0^{2\pi} \log |f(re^{i\theta})|\, d\theta$$

where a zero of order p is counted p times. (This formula connects the modulus of a function with the moduli of the zeros.)

Proof. First write the formula in another way. Let $n(x)$ denote the number of zeros of $f(z)$ for $|z| \leq x$. Then if $r_n \leq r \leq r_{n+1}$,

$$\log \frac{r^n}{r_1 r_2 \cdots r_n} = n \log r - \sum_{m=1}^{n} \log r_m$$

$$= \sum_{m=1}^{n-1} m(\log r_{m+1} - \log r_m) + n(\log r - \log r_n)$$

$$= \sum_{m=1}^{n-1} m \int_{r_m}^{r_{m+1}} \frac{dx}{x} + n \int_{r_n}^{r} \frac{dx}{x}.$$

NOTE. For $r_m \le x < r_{m+1}$, $m = n(x)$ and $n = n(x)$ for $r_n \le x < r$. Thus

$$\log \frac{r^n}{r_1 \cdots r_n} = \int_0^r \frac{n(x)}{x}\, dx = \sum_{m=1}^{n-1} m \int_{r_m}^{r_{m+1}} \frac{dx}{x} + n \int_{r_n}^{r} \frac{dx}{x}.$$

{Check:

$$\int_0^r \frac{n(x)}{x}\, dx = \int_0^{r_1} 0 \cdot \frac{dx}{x} + \int_{r_1}^{r_2} 1 \cdot \frac{dx}{x} + \cdots + \int_{r_n}^{r} n \cdot \frac{dx}{x}.$$

$$= 1 \cdot \int_{r_1}^{r_2} \frac{dx}{x} + \cdots + (n-1) \int_{r_{n-1}}^{r_n} \frac{dx}{x} + n \int_{r_n}^{r} \frac{dx}{x}.\}$$

Thus we require to prove

$$\int_0^r \frac{n(x)}{x}\, dx = \frac{1}{2\pi} \int_0^{2\pi} \log |f(re^{i\theta})|\, d\theta - \log |f(0)|.$$

Clearly both sides of the formula are equal for $r = 0$.

If $f(z)$ has no zero on $|z| = r$,

$$n(r) = \frac{1}{2\pi i} \int_{|z|=r} \frac{f'(z)}{f(z)}\, dz = \frac{1}{2\pi} \int_0^{2\pi} \frac{f'(re^{i\theta})}{f(re^{i\theta})}\, re^{i\theta}\, d\theta. \qquad (1)$$

We clearly cannot divide by r, integrate with respect to r, and take real parts, by virtue of infinities of the integrand. In an interval between the moduli of two zeros r_n, r_{n+1} each side of Jensen's formula has a continuous derivative. We examine the assertion of the theorem. The derivative of the left-hand side

$$\frac{d}{dr}\left\{\log \frac{r^n |f(0)|}{r_1 \cdots r_n}\right\} = \frac{n}{r}.$$

The derivative of the right-hand side is

$$\frac{1}{2\pi} \int_0^{2\pi} \frac{d}{dr} \{\log |f(re^{i\theta})|\}\, d\theta = \frac{1}{4\pi} \int_0^{2\pi} \frac{d}{dr} \{\log f(re^{i\theta}) + \log \bar{f}(re^{-i\theta})\}\, d\theta$$

(since $\log |f|^2 = 2 \log |f| = \log f \cdot \bar{f}$)

$$= \frac{1}{4\pi} \int_0^{2\pi} \left\{\frac{f'(re^{i\theta})}{f(re^{i\theta})}\, e^{i\theta} + \frac{\bar{f}'(re^{-i\theta})}{\bar{f}(re^{-i\theta})}\, e^{-i\theta}\right\} d\theta$$

$$= \mathscr{R}\left\{\frac{1}{2\pi} \int_0^{2\pi} \frac{f'(re^{i\theta})}{f(re^{i\theta})}\, e^{i\theta}\, d\theta\right\}.$$

This result equals $n(r)/r = n/r$ by (1) also.

Hence the two derivatives are equal in any such interval. Hence the two sides of Jensen's formula differ by a constant in any such interval. Since both sides of Jensen's formula are equal for $r = 0$, the constant is 0 (in that particular interval). Therefore it is sufficient to prove that each side (of Jensen's formula) is continuous when r passes through a value r_n. This is clearly true for the left-hand side. We consider the right-hand side. It is sufficient to assume that there is one zero of modulus r_n and amplitude zero. Thus $z_n = r_n$ and $f(z) = (z_n - z)\phi(z)$, where $\phi(z)$ is analytic and nonzero in the neighborhood of $z = z_n$. Consequently

$$\begin{aligned}\log|f(z)| &= \log|r_n - re^{i\theta}| + \log|\phi(z)| \\ &= \log\left|1 - \frac{r}{r_n}e^{i\theta}\right| + \psi(r, \theta)\end{aligned}$$

where $\psi(r, \theta) \neq 0$ and continuous in the neighborhood of $z = z_n$. Since we are considering a neighborhood of $z = z_n$, i.e., $r = r_n$, let $r/r_n < 2$ and $|\theta| < \pi$. Then

$$\left|1 - \frac{r}{r_n}e^{i\theta}\right|^2 < 9, \quad \text{since} \quad \left|1 - \frac{r}{r_n}e^{i\theta}\right| \leq 1 + \left|\frac{r}{r_n}\right| < 3$$

and therefore

$$\begin{aligned}9 > \left|1 - \frac{r}{r_n}e^{i\theta}\right|^2 &= 1 - \frac{2r}{r_n}\cos\theta + \left(\frac{r}{r_n}\right)^2 \\ &= \sin^2\theta + \left(\cos\theta - \frac{r}{r_n}\right)^2 \geq \sin^2\theta.\end{aligned}$$

Further, in the neighborhood of the real zero $z = r_n$, we require to consider a range of integration about $\theta = 0$, viz $[-\delta, \delta]$ as $\delta \to 0$. Thus

$$\begin{aligned}\left|\int_{-\delta}^{\delta} \log\left|1 - \frac{r}{r_n}e^{i\theta}\right| d\theta\right| &\leq \int_{-\delta}^{\delta} \left|\log\left|1 - \frac{r}{r_n}e^{i\theta}\right|\right| d\theta \\ &< \int_{-\delta}^{\delta} \left\{\log 3 + \left|\log|\sin\theta|\right|\right\} d\theta.\end{aligned}$$

By writing

$$\left(1 - \frac{r}{r_n}e^{i\theta}\right) = \alpha,$$

we have

$$|\log|\alpha|| < \log 3$$

and

$$\log |\alpha| \geq \log |\sin \theta|,$$

thus

$$|\log |\alpha|| \leq |\log |\sin \theta||$$

and hence

$$|\log |\alpha|| < \log 3 + |\log |\sin \theta||.$$

Also, for $0 < \theta < \pi/2$, we have $\sin \theta > 2\theta/\pi$. Thus

$$\log \sin \theta = \log |\sin \theta| > B + \log \theta,$$

(since $\sin \theta$ is positive) and

$$|\log |\sin \theta|| < |B + \log \theta| \leq C + |\log \theta| \qquad (\log |\sin \theta| < 0).$$

Hence (see Fig. 1)

$$\begin{aligned}\left|\int_{-\delta}^{\delta} \log \left|1 - \frac{r}{r_n} e^{i\theta}\right| d\theta\right| &< \int_{-\delta}^{\delta} \{A + |\log |\theta||\}\, d\theta \\ &= 2A\delta - 2\int_{\varepsilon}^{\delta} \log \theta \, d\theta, \qquad \varepsilon \to 0 \\ &= 2A\delta - 2\left\{\theta \log \theta - \theta\right\}_{\varepsilon}^{\delta} \\ &= 2A\delta - 2\delta \log \delta + 2\delta, \qquad \varepsilon \log \varepsilon \to 0, \\ &\to 0 \qquad \text{since} \quad \delta \log \delta \to 0.\end{aligned}$$

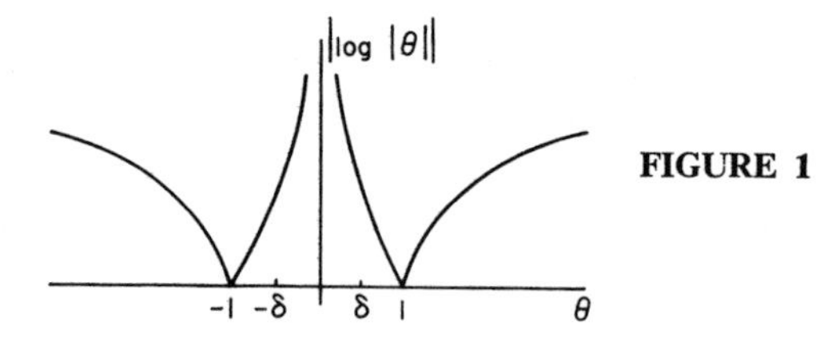

FIGURE 1

Thus a contribution to the integral in the neighborhood of zero is arbitrarily small; thus the whole integral is continuous and Jensen's theorem is established. □

The theorem may be extended to functions with poles as well as zeros.

Let $f(z)$ satisfy the same conditions of the original theorem and let it have zeros $a_1, \ldots, a_m$ and poles $b_1, \ldots, b_n$ with moduli less than or equal to r. Then

$$\log\left\{\left|\frac{b_1 \cdots b_n}{a_1 \cdots a_n} f(0)\right| r^{m-n}\right\} = \frac{1}{2\pi} \int_0^{2\pi} \log | f(re^{i\theta}) | \, d\theta,$$

for let

$$f(z) = g(z) \Big/ \left(1 - \frac{z}{b_1}\right)\left(1 - \frac{z}{b_2}\right) \cdots \left(1 - \frac{z}{b_n}\right) = \frac{g(z)}{h(z)}.$$

Since $g(z)$ has zeros $a_1, \ldots, a_m$, $f(0) = g(0)$,

$$\log \frac{r^m g(0)}{| a_1 a_2 \cdots a_m |} = \frac{1}{2\pi} \int_0^{2\pi} \log | g(re^{i\theta}) | \, d\theta \qquad (1)$$

and

$$\log \frac{r^n h(0)}{| b_1 b_2 \cdots b_n |} = \log \frac{r^n}{| b_1 b_2 \cdots b_n |} = \frac{1}{2\pi} \int_0^{2\pi} \log | h(re^{i\theta}) | \, d\theta \qquad (2)$$

The result follows by subtracting (2) from (1).

3.4 THE POISSON–JENSEN FORMULA

Let $f(z)$ have zeros at $a_1, a_2, \ldots, a_m$ and poles at $b_1, b_2, \ldots, b_n$, inside the circle $| z | \leq R$ and let $f(z)$ be analytic elsewhere inside and on the circle. Then

$$\log | f(re^{i\theta}) | = \frac{1}{2\pi} \int_0^{2\pi} \frac{R^2 - r^2}{R^2 - 2Rr\cos(\theta - \phi) + r^2} \log | f(Re^{i\phi}) | \, d\phi$$
$$- \sum_{\mu=1}^{m} \log\left| \frac{R^2 - \bar{a}_\mu re^{i\theta}}{R(re^{i\theta} - a_\mu)} \right| + \sum_{\nu=1}^{n} \log\left| \frac{R^2 - \bar{b}_\nu re^{i\theta}}{R(re^{i\theta} - b_\nu)} \right|.$$

This contains Poisson's and Jensen's formulas as special cases. For no zeros or poles, we have Poisson's formula for the real part of $\log f(z)$, viz.,

$$\log | f(re^{i\theta}) | = \frac{1}{2\pi} \int_0^{2\pi} \frac{R^2 - r^2}{R^2 - 2Rr\cos(\theta - \phi) + r^2} \log | f(Re^{i\phi}) | \, d\phi.$$

For $r = 0$,

$$\log | f(0) | = \frac{1}{2\pi} \int_0^{2\pi} \log | f(Re^{i\phi}) | \, d\phi - \log\left\{\left| \frac{b_1 b_2 \cdots b_n}{a_1 a_2 \cdots a_m} \right| R^{m-n}\right\}.$$

I. Let $f(z) = z - a$, $|a| < R$. We require to prove that

$$\log |re^{i\theta} - a| = \frac{1}{2\pi}\int_0^{2\pi} \frac{R^2 - r^2}{R^2 - 2Rr\cos(\theta - \phi) + r^2} \log |Re^{i\phi} - a|\, d\phi$$

$$- \log\left|\frac{R^2 - \bar{a}re^{i\theta}}{R(re^{i\theta} - a)}\right|,$$

i.e.,

$$\log\left|R - \frac{\bar{a}re^{i\theta}}{R}\right| = \frac{1}{2\pi}\int_0^{2\pi} \frac{R^2 - r^2}{R^2 - 2Rr\cos(\theta - \phi) + r^2} \log |Re^{i\phi} - a|\, d\phi;$$

and this is true by applying Poisson's formula to the real part of $\log(R - (\bar{a}z/R))$ which is analytic for $|z| \leq R$. Since $U(r, \theta)$ corresponds to $\log|R - (\bar{a}re^{i\theta}/R)|$, $U(R, \phi)$ corresponds to $\log|R - \bar{a}\, e^{i\phi}|$, $a = \alpha + i\beta$, and

$$U(R, \phi) = \tfrac{1}{2}\log\{(R - \alpha\cos\phi - \beta\sin\phi)^2 + (\beta\cos\phi - \alpha\sin\phi)^2\},$$
$$\log|Re^{i\phi} - a| = \tfrac{1}{2}\log\{(R\cos\phi - \alpha)^2 + (R\sin\phi - \beta)^2\},$$

thus

$$\log|R - \bar{a}e^{i\phi}| = \log|Re^{i\phi} - a|.$$

II. If $f(z) = 1/(z - b)$, the Poisson–Jensen formula is equivalent to Poisson's formula for the real part of $\log\big(R - (\bar{b}z/R)\big)$.

III. If $f(z)$ is analytic with no poles or zeros in $|z| \leq R$, the formula is Poisson's formula for the real part of $\log f(z)$.

We now add all these cases to obtain the Poisson–Jensen formula.

3.5 A theorem is included now, which has considerable application in gap and density theorems. Jensen's theorem applies to a circular region and the following theorem is similar but applies to a half-plane.

3.5.1 Carleman's Theorem. Let $f(z)$ be analytic for

$$|z| \geq \varrho, \quad -\pi/2 \leq \arg z \leq \pi/2$$

and let it have zeros $r_1e^{i\theta_1}, r_2e^{i\theta_2}, \ldots, r_ne^{i\theta_n}$, inside the contour consisting of semicircles $|z| = \varrho$, $|z| = R$, $-\pi/2 \leq \arg z \leq \pi/2$ and the parts of the imaginary axis joining them. Let $f(z)$ have no zeros on the contour.

Then

$$\sum_{\mu=1}^{n}\left(\frac{1}{r_\mu}-\frac{r_\mu}{R^2}\right)\cos\theta_\mu=\frac{1}{\pi R}\int_{-\pi/2}^{\pi/2}\log|f(Re^{i\theta})|\cos\theta\,d\theta$$

$$+\frac{1}{2\pi}\int_\varrho^R\left(\frac{1}{y^2}-\frac{1}{R^2}\right)\log\{|f(iy)|\;|f(-iy)|\}\,dy+O(1),$$

Where $O(1)$ denotes a function of ϱ, R which for fixed ϱ is bounded as $R\to\infty$.

Proof. Consider

$$I=\frac{1}{2\pi i}\int_C\log f(z)\left(\frac{1}{z^2}+\frac{1}{R^2}\right)dz,$$

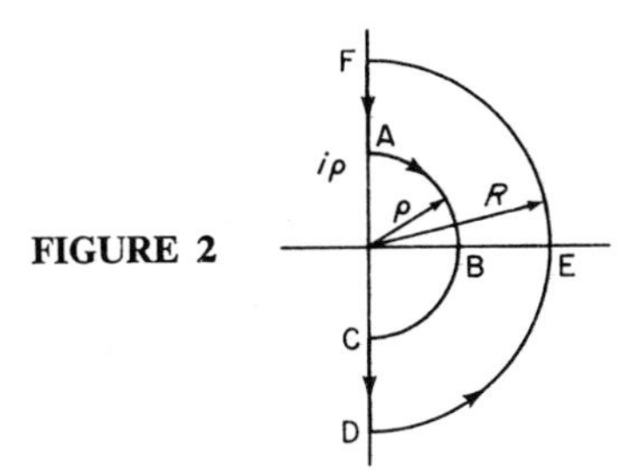

FIGURE 2

taken around the contour C: $ABCDEF$ (Fig. 2) starting at $z=i\varrho$ and a fixed determination of the log starting at $z=i\varrho$ and thereafter varying continuously, $\int_{C_\varrho}$ is bounded. On the negative imaginary axis $z=-iy$ and the contribution to I is

$$I_1=\frac{1}{2\pi}\int_\varrho^R\log f(-iy)\left\{\frac{1}{y^2}-\frac{1}{R^2}\right\}dy.$$

On the large semicircle, $z=Re^{i\theta}$ and we obtain

$$\frac{1}{2\pi i}\int_{-\pi/2}^{\pi/2}\log f(Re^{i\theta})\left\{\frac{e^{-2i\theta}}{R^2}+\frac{1}{R^2}\right\}iRe^{i\theta}\,d\theta=\frac{1}{\pi R}\int_{-\pi/2}^{\pi/2}\log f(Re^{i\theta})\cos\theta\,d\theta$$

since $e^{i\theta}+e^{-i\theta}=2\cos\theta$. Along the positive imaginary axis, the contribution to I becomes

$$I_2=\frac{1}{2\pi}\int_\varrho^R\log f(iy)\left(\frac{1}{y^2}-\frac{1}{R^2}\right)dy.$$

Adding and taking the real part, we obtain the right-hand side of Carleman's theorem, viz.,

$$\frac{1}{\pi R}\int_{-\pi/2}^{\pi/2}\log|f(Re^{i\theta})|\cos\theta\,d\theta$$
$$+\frac{1}{2\pi}\int_{\varrho}^{R}\left\{\frac{1}{y^2}-\frac{1}{R^2}\right\}\log\{|f(iy)f(-iy)|\}\,dy+O(1).$$

Also, integrating I by parts, we obtain

$$I=\frac{1}{2\pi i}\left\{\log f(z)\left(\frac{z}{R^2}-\frac{1}{z}\right)\right\}_C+\frac{1}{2\pi i}\int_C\frac{f'(z)}{f(z)}\left(\frac{1}{z}-\frac{z}{R^2}\right)dz.$$

In moving around the contour starting at $z=i\varrho$ and initial value of the log, the log increases by $2n\pi i$ finishing at $z=i\varrho$. Hence

$$\log f(z)\left(\frac{z}{R^2}-\frac{1}{z}\right)\Big]_C=\{\log f(i\varrho)+2n\pi i\}\left(\frac{i\varrho}{R^2}-\frac{1}{i\varrho}\right)$$
$$-\log f(i\varrho)\left(\frac{i\varrho}{R^2}-\frac{1}{i\varrho}\right)$$
$$=2n\pi i\left(\frac{i\varrho}{R^2}-\frac{1}{i\varrho}\right),$$

where n is the number of zeros of f in the contour. Thus the integrated term becomes purely imaginary, viz.,

$$n\left(\frac{i\varrho}{R^2}-\frac{1}{i\varrho}\right).$$

By the corollary to Theorem 2.3.1 we have

$$\mathscr{R}\{I\}=\mathscr{R}\left\{\frac{1}{2\pi i}\int_C\frac{f'(z)}{f(z)}\left(\frac{1}{z}-\frac{z}{R^2}\right)dz\right\}$$
$$=\mathscr{R}\left\{\sum_{\mu=1}^{n}\left(\frac{1}{r_\mu e^{i\theta_\mu}}-\frac{r_\mu e^{i\theta_\mu}}{R^2}\right)\right\}$$
$$=\sum_{\mu=1}^{n}\left(\frac{1}{r_\mu}-\frac{r_\mu}{R^2}\right)\cos\theta_\mu,$$

hence the result. □

Note. In the proof of Carleman's theorem, $\log f(z)$ is not a single-valued analytic function in the region and we do not apply Cauchy's theorem to any function involving $\log f$. (If we did, we should have to make our region simply-connected by a canal Γ, as in Fig. 3, joining all zeros of f to the boundary C. We simply choose one of the possible values of $\log f$ at $z = i\varrho$, follow it around C, and use it in

$$I_\gamma + I_{C-\gamma} = I = \frac{1}{2\pi i} \int_C \log f(z) \cdot \left(\frac{1}{z^2} + \frac{1}{R^2}\right) dz.$$

The first part shows that there is no ambiguity in $\mathscr{R}\{I_{C-\gamma}\}$ in spite of the ambiguity in $\log f$ and therefore in I. Then $\mathscr{R}\{I_{C-\gamma}\}$ gives the main term on the right-hand side of the theorem.

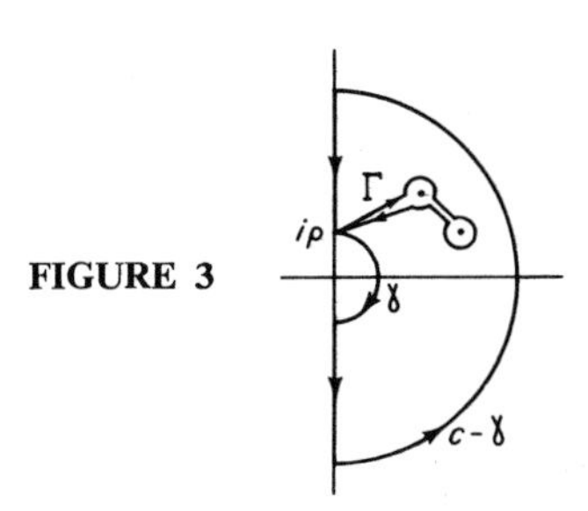

FIGURE 3

It also remarks that once the branch of $\log f$ at $i\varrho$ is chosen, I_γ, and so also $\mathscr{R}\{I_\gamma\}$, is bounded. (This is not "as $\varrho \to 0$" but "as $R \to \infty$." In fact ϱ is not varied.) Thus $\mathscr{R}\{I_\gamma\}$ is the $O(1)$ term. Integration by parts then replaces I by "integrated terms" and the integral

$$\frac{1}{2\pi i} \int_C \frac{f'(z)}{f(z)} \left(\frac{1}{z} - \frac{z}{R^2}\right) dz.$$

To this new integral the theory of residues is applied.

The proof that $\mathscr{R}\{I_\gamma\}$ is $O(1)$ is via $I_\gamma = o(1)$. For this it is necessary that we fix a value of $\log f$ around the whole of γ and that no part of γ uses a $\log f$ with some large addition, $+\, 2n\pi i$. So we fix a value of the $\log f$ at $i\varrho$ or at some fixed point iy_0 above $i\varrho$, and deduce $\log f$ from there onward counterclockwise. "Fix" means determine independently of R.

To fix at $-i\varrho$ would have the same effect, since

$$\mathscr{R}\left\{\frac{1}{2\pi i} \int_\gamma (2\pi i n)\left(\frac{1}{z^2} + \frac{1}{R^2}\right) dz\right\} = n \cdot \mathscr{R}\left[\frac{z}{R^2} - \frac{1}{z}\right]_{i\varrho}^{-i\varrho} = 0.$$

However, to fix partway along γ, e.g. at ϱ, would fail to prove that the unknown correction is $O(1)$. Both I_γ and the "integrated terms" would give contributions of order n and we would have to prove that they cancel out. Actually $O(1) = A + (B/R^2)$, where A, B are independent of R. For applications of this theorem, see Titchmarsh[†] and Levinson.[‡] For example, it is now a very simple matter to show that if $f(z)$ is analytic and bounded in the right half-plane and if the zeros in the right half-plane are $r_1 e^{i\theta_1}$, $r_2 e^{i\theta_2}$, ..., then the series

$$\sum_{\mu=1}^{\infty} \frac{\cos\theta_\mu}{r_\mu} \quad \text{converges}.$$

We continue with several theorems giving bounds for the absolute value of a function in terms of various parameters. It is possible in some cases to "tighten-up" the theorems, to improve bounds, or to generalize, however, since we only require the following forms of the theorems we leave it to the reader to consult the literature for further improvements.

3.6 Theorem. (Schwarz's lemma). If $f(z)$ is analytic for $|z| < R$, continuous for $|z| \leq R$, and $f(0) = 0$, then

$$|f(re^{i\theta})| \leq rM/R, \qquad 0 \leq r \leq R,$$

where $M = \max_{|z|=R} |f(z)|$. Equality only occurs when $f(z) = zMe^{i\gamma}/R$, and γ is a real constant.

Proof. Let

$$g(z) = \frac{f(z)}{z} \qquad \text{for} \quad 0 < |z| \leq R$$

and

$$g(0) = f'(0).$$

Then $g(z)$ is analytic for $0 < |z| < R$ and continuous for $0 \leq |z| \leq R$. Thus $\max |g(z)|$ occurs either at $z = 0$ or $|z| = R$.

$$|g(0)| = |f'(0)| \leq M/R \qquad \text{for} \quad z = 0,$$

$$|g(z)| = |f(z)|/|z| \leq M/R \qquad \text{for} \quad |z| = R.$$

Hence

$$|g(z)| \leq M/R \qquad \text{for} \quad |z| \leq R,$$

[†] E. C. Titchmarsh, "The Theory of Functions." Oxford Univ. Press, London and New York, 1939.

[‡] N. Levinson, "Gap and Density Theorems" (AMS Colloq. Publ.), Amer. Math. Soc., Providence, Rhode Island, 1940 (reprinted 1963).

equality holding when $g(z)$ is a constant $Me^{i\gamma}/R$. Therefore

$$|f(re^{i\theta})| \leq rM/R \qquad \text{for} \quad r \leq R,$$

equality holding only when $f(z) = zMe^{i\gamma}/R$. □

3.7 A THEOREM OF BOREL AND CARATHÉODORY

The following result enables us to deduce an upper bound for the modulus of a function on $|z| = r$, from bounds for its real or imaginary parts on a larger concentric circle $|z| = R$.

3.7.1 Theorem. Let $f(z)$ be analytic for $|z| \leq R$ and let $M(r)$ and $A(r)$ denote the max $|f(z)|$ and max $\mathscr{R}\{f(z)\}$ on $|z| = r$. Then for $0 < r < R$,

$$M(r) \leq \frac{2r}{R-r} A(R) + \frac{R+r}{R-r} |f(0)| < \frac{R+r}{R-r} \{A(R) + |f(0)|\}.$$

Proof. The result is clearly true for $f(z) =$ constant. For $f(z)$ nonconstant suppose $f(0) = 0$. Then $A(R) > A(0) = 0$. Since by the Poisson–Jensen formula

$$U(0) = \frac{1}{2\pi} \int_0^{2\pi} U(R, \phi)\, d\phi,$$

and if we let $A(R) = \max_\phi U(R, \phi)$, then $U(R, \phi) \leq A(R)$, $\forall \phi$ and

$$U(0) = \frac{1}{2\pi} \int_0^{2\pi} U(R, \phi)\, d\phi \leq \frac{1}{2\pi} \int_0^{2\pi} A(R)\, d\phi = A(R).$$

Thus $U(0) = A(R)$ implies $U(R, \phi) = A(R)$, $\forall \phi$ and $A(0) \leq U(0) \leq A(R)$ unless the function is a constant.

Let

$$\phi(z) = \frac{f(z)}{2A(R) - f(z)}.$$

Then $\phi(z)$ is analytic for $|z| \leq R$ since the real part of the denominator does not vanish. Now $\phi(0) = 0$ and if $f(z) = u + iv$,

$$|\phi(z)|^2 = \frac{u^2 + v^2}{\{2A(R) - u\}^2 + v^2} \leq 1$$

$$\text{since} \quad -2A(R) + u \leq u \leq 2A(R) - u.$$

Also, Schwarz's lemma gives

$$|\phi(z)| \leq r/R$$

since $|\phi(z)|$ is bounded by 1. Thus

$$|f(z)| = \frac{|2A(R)\phi(z)|}{|1 + \phi(z)|} \leq \frac{2A(R) \cdot r}{R - r}$$

(use $|R + R\phi| \geq R - R|\phi|$). Thus we obtain the result for $f(0) = 0$. If $f(0) \neq 0$, apply the result to $f(z) - f(0)$ and

$$\begin{aligned} |f(z) - f(0)| &\leq \frac{2r}{R - r} \cdot \max_{|z|=R} \{\mathscr{R}(f(z) - f(0))\} \\ &\leq \frac{2r}{R - r} \{A(R) + |f(0)|\} \end{aligned}$$

since $f(0)$ may be negative. Adding $|f(0)|$ to both sides, we have that

$$|f(z)| \leq \frac{2r}{R - r} A(R) + \frac{R + r}{R - r} |f(0)|. \quad \square$$

We also show that

$$\max_{|z|=r} |f^{(n)}(z)| \leq \frac{2^{n+2} n! R}{(R - r)^{n+1}} \{A(R) + |f(0)|\}$$

since

$$f^{(n)}(z) = \frac{n!}{2\pi i} \int_C \frac{f(\omega) - f(0)}{(\omega - z)^{n+1}} d\omega \tag{1}$$

where C is the circle

$$|\omega - z| = \delta = \tfrac{1}{2}(R - r)$$

and

$$|\omega| \leq r + \tfrac{1}{2}(R - r) = \tfrac{1}{2}(R + r)$$

which ensures C is inside $|z| = R$ (Fig. 4). The previous theorem gives

$$\max |f(z) - f(0)| \leq \frac{R + \frac{1}{2}(R + r)}{R - \frac{1}{2}(R + r)} \{A(R) + |f(0)|\}$$

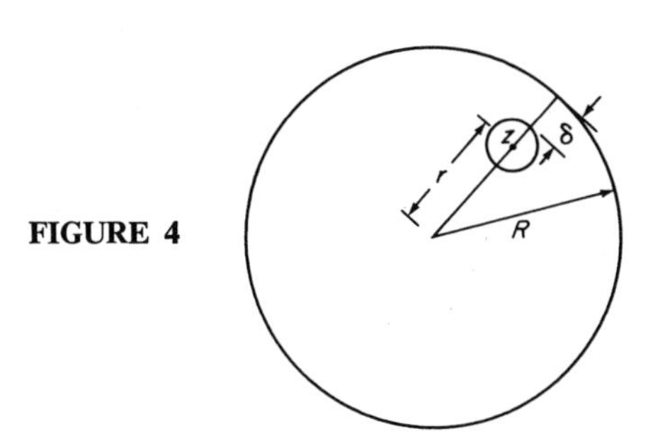

FIGURE 4

and since $r < R$,

$$\max |f(z) - f(0)| < \frac{4R}{R-r}\{A(R) + |f(0)|\}.$$

Thus from Eq. (1)

$$|f^{(n)}(z)| \leq \frac{n!4R}{\delta^n(R-r)}\{A(R) + |f(0)|\} = \frac{2^{n+2}n!R}{(R-r)^{n+1}}\{A(R) + |f(0)|\}.$$

CHAPTER IV

INFINITE PRODUCT REPRESENTATION: ORDER AND TYPE

A study will now be made of the infinite product representation of an entire function. Several distinctions between functions with an infinite number of zeros and functions with a finite number of zeros will emerge. The concept of "order" will be dealt with in more detail, leading to a further object for studying these functions, namely "type." We start with a well-known and very illuminating theorem of Weierstrass.

4.1 THE WEIERSTRASS FACTORIZATION THEOREM

Since an entire function is an analytic function with no singularities except at ∞, consider a polynomial $f(z)$ with zeros at $z_1, z_2, \ldots, z_n$. The polynomial can then be factorized as

$$f(z) = f(0)\left(1 - \frac{z}{z_1}\right)\left(1 - \frac{z}{z_2}\right)\cdots\left(1 - \frac{z}{z_n}\right)$$

Thus given any finite set of points $\{z_1, z_2, \ldots, z_n\}$, there always exists an entire function with zeros $z_1, z_2, \ldots, z_n$. Consider now an infinite sequence of points $z_1, z_2, \ldots, z_n, \ldots$ such that $0 < |z_1| \leq |z_2| \leq \cdots \leq |z_n| \leq \cdots$ and whose sole limit point in ∞, i.e., $\lim_{n\to\infty} |z_n| = \infty$.

We must construct an entire function which vanishes at these points and nowhere alse $\{\prod_{n=1}^{\infty}(1-(z/z_n))$ unfortunately may diverge$\}$. This is, in effect, accomplished by Weierstrass's factorization theorem, which demonstrates the construction and existence of such a function. We prove a form of the theorem stated as follows:

4.1.1 Theorem. If $\{z_n\}$ is an arbitrary sequence of complex numbers different from zero and whose sole limit point is ∞ and if m is a nonnegative integer, then $\exists$ an entire function $G(z)$ having roots at the points z_1, z_2, ... (and these points only) and a root of multiplicity m at the point zero. Further, $G(z)$ can be defined by the absolutely uniformly convergent product

$$G(z) = e^{g(z)} z^m \prod_{n=1}^{\infty} \left(1 - \frac{z}{z_n}\right) e^{Q_\nu(z/z_n)}, \qquad \text{where } e^{g(z)}$$

is an arbitrary entire function and $Q_\nu(z)$ is a polynomial such that

$$Q_\nu(z) = z + \frac{z^2}{2} + \cdots + \frac{z^\nu}{\nu}.$$

The nonnegative integer ν has the property that the series $\sum_{n=1}^{\infty} 1/|z_n|^{\nu+1}$ converges uniformly in the whole plane.

Proof. Consider $(1-(z/z_n))e^{Q_\nu(z/z_n)}$ where $Q_\nu(z)$ is a polynomial of degree ν. This is an entire function which vanishes for $z = z_n$. Since

$$\left(1 - \frac{z}{z_n}\right) e^{Q_\nu(z/z_n)} = e^{Q_\nu(z/z_n)+\log(1-z/z_n)}$$

where $|z/z_n| < 1$,

$$\left(1 - \frac{z}{z_n}\right) \exp Q_\nu(z/z_n) = \exp\left(-\frac{z}{z_n} - \frac{z^2}{2z_n{}^2} - \cdots + Q_\nu\left(\frac{z}{z_n}\right)\right)$$

and we choose $Q_\nu(z/z_n) = (z/z_n) + \cdots + (z^\nu/\nu z_n{}^\nu)$. Then

$$\left(1 - \frac{z}{z_n}\right) \exp Q_\nu\left(\frac{z}{z_n}\right) = \exp\left(-\left(\frac{z}{z_n}\right)^{\nu+1} \frac{1}{\nu+1} - \cdots\right) = 1 + U_n(z).$$

We are required to determine ν such that $\prod_{n=1}^{\infty}(1-(z/z_n))e^{Q_\nu(z/z_n)}$ is absolutely and uniformly convergent for $|z| < R$ and arbitrarily large R. Choose $R > 1$ and α such that $0 < \alpha < 1$. Then $\exists$ a positive integer q such that $|z_q| \leq R/\alpha$ and $|z_{q+1}| > R/\alpha$. Thus we see that the partial product $\prod_{n=1}^{q}(1 - (z/z_n))e^{Q_\nu(z/z_n)}$ is trivially an entire function. Consider the remainder $\prod_{n=q+1}^{\infty}(1 - (z/z_n))e^{Q_\nu(z/z_n)}$ with $|z| \leq R$. For $n > q$, $|z_n| > R/\alpha$, i.e., $|z/z_n| < \alpha < 1$. We now estimate each factor $\{1 + U_n(z)\}$ in the product for $n > q$

$$|U_n(z)| = \left|\exp\left(-\frac{1}{\nu+1}\left(\frac{z}{z_n}\right)^{\nu+1} - \cdots\right) - 1\right|$$

$$\leq \exp\left|\frac{1}{\nu+1}\left(\frac{z}{z_n}\right)^{\nu+1} + \cdots\right| - 1$$

since

$$|e^m - 1| \leq |m| + \frac{|m|^2}{2!} + \cdots = e^{|m|} - 1.$$

Thus

$$|U_n(z)| < \exp\left(\left|\frac{z}{z_n}\right|^{\nu+1}\left(1 + \left|\frac{z}{z_n}\right| + \cdots\right)\right) - 1$$

$$< \exp\left(\left|\frac{z}{z_n}\right|^{\nu+1}(1 + \alpha + \alpha^2 + \cdots)\right) - 1$$

$$= \exp\left(\frac{1}{1-\alpha}\left|\frac{z}{z_n}\right|^{\nu+1}\right) - 1$$

$$\leq \frac{1}{1-\alpha}\left|\frac{z}{z_n}\right|^{\nu+1}\exp\left(\frac{1}{1-\alpha}\left|\frac{z}{z_n}\right|^{\nu+1}\right),$$

since $e^m - 1 \leq me^m$. Hence

$$|U_n(z)| < \frac{e^{1/1-\alpha}}{1-\alpha}\left|\frac{z}{z_n}\right|^{\nu+1}.$$

Two cases arise:

I. either $\exists$ a positive integer p such that $\sum_{n=1}^{\infty} 1/|z_n|^p < \infty$, or

II. there does not exist such an integer.

Case I. Take $\nu = p - 1$ and

$$|U_n(z)| < \frac{R^p}{|z_n|^p} \cdot \frac{e^{1/1-\alpha}}{1-\alpha}$$

since $|z| \leq R$, and $\sum_{(n)} |U_n(z)| < \infty$ for $|z| \leq R$. Thus, for $|z| \leq R$, $\prod_{n=q+1}^{\infty}(1 + U_n(z))$ converges absolutely and uniformly.

CASE II. Take $\nu = n - 1$ so that

$$| U_n(z) | < \frac{e^{1/1-\alpha}}{1-\alpha} \left| \frac{z}{z_n} \right|^n \quad \text{for} \quad n > q, \quad \left| \frac{z}{z_n} \right| < 1,$$

$$| z | \leq R \quad , \quad | z_n | \to \infty .$$

Thus by the root test $\sum_{(n)} | U_n(z) | < \infty$ with the same result as before. Thus $\prod_{n=1}^{\infty}(1 - (z/z_n))e^{Q_\nu(z/z_n)}$ is analytic in $| z | \leq R$, and since R is arbitrary and $\nu \neq \nu(R)$ the product represents an entire function. If, further, $z = 0$ is a zero of order m of $G(z)$, then $G(z)/z^m G_1(z)$ has no zeros and equals $e^{g(z)}$, say, where $G_1(z)$ is the above infinite product, and

$$G(z) = e^{g(z)} z^m \prod_{n=1}^{\infty} \left(1 - \frac{z}{z_n}\right) e^{Q_\nu(z/z_n)}.$$

Here $g(z)$ is an arbitrary entire function. □

If $G(z)$ is subject to further restrictions, it should be possible to say more about $g(z)$. The factor $e^{Q_\nu(z/z_n)}$ is clearly inserted to produce convergence of the infinite product. The factors

$$\left(1 - \frac{z}{z_n}\right) \exp\left(\frac{z}{z_n} + \cdots + \frac{1}{\nu}\left(\frac{z}{z_n}\right)^\nu\right)$$

are called "primary factors" and the product formed with these factors is called the "canonical product," provided that ν is the smallest integer for which $\sum_{(n)} 1/| z_n |^{\nu+1}$ converges.

4.2 ORDER OF AN ENTIRE FUNCTION

We define *order* of an entire function such that f is of finite order, if there exists a constant λ such that $| f(z) | < \exp r^\lambda$ for $| z | = r > r_0$, and for nonconstant f of finite order, $\lambda > 0$. If the inequality is true for a certain λ, then it is true for $\lambda' > \lambda$, thus there exists an infinity of λ's > 0 satisfying this inequality.

The lower bound of these λ's is called the "order of f." Denote this lower bound by ϱ. Then given $\varepsilon > 0$, $\exists r_0$ such that

$$| f(z) | < \exp r^{\varrho+\varepsilon} \quad \text{for} \quad | z | = r > r_0 .$$

This implies

$$M(r) = \max_{|z|=r} | f(z) | < \exp r^{\varrho+\varepsilon} \quad \text{for} \quad r > r_0 ,$$

while $M(r) > \exp r^{\varrho-\varepsilon}$ for an infinite number of $r \to +\infty$. Thus

$$\frac{\log\log M(r)}{\log r} < \varrho + \varepsilon \qquad \text{and} \qquad \frac{\log\log M(r)}{\log r} > \varrho - \varepsilon$$

for an infinite number of $r \to +\infty$, hence

$$\varrho = \overline{\lim_{r\to\infty}} \frac{\log\log M(r)}{\log r}.$$

Alternatively, $f(z)$ is of finite order, if $\exists A > 0$ such that as $|z| = r \to \infty$

$$|f(z)| = O(\exp r^A).$$

Thus $f(z)$ is of order ϱ if

$$|f(z)| = O(\exp r^{\varrho+\varepsilon})$$

for every $\varepsilon > 0$ but not for any negative ε. The constant implied in O depends in general on ε, otherwise ε could be replaced by zero in the formula. A few theorems will now be proved using the above expression for the order.

4.2.1 Theorem. If $f_1(z)$ and $f_2(z)$ are entire functions of order ϱ_1 and ϱ_2, respectively, and if $\varrho_1 < \varrho_2$, then the order of $F(z) = f_1(z) + f_2(z)$ is equal to ϱ_2.

Proof. We suppose ϱ_2 is finite. Since

$$\begin{aligned} M(r;\, f_1 + f_2) &\leq M(r;\, f_1) + M(r;\, f_2) \\ &\leq \exp r^{\varrho_1+\varepsilon} + \exp r^{\varrho_2+\varepsilon} \\ &< 2\exp r^{\varrho_2+\varepsilon} \qquad \text{for} \quad r > r_0(\varepsilon), \end{aligned}$$

we have that $\varrho \leq \varrho_2 + \varepsilon$ and hence $\varrho \leq \varrho_2$. On the other hand $\exists$ a sequence of numbers $r_n \to \infty$ such that $M(r_n;\, f_2) > \exp r_n^{\varrho_2-\varepsilon}$. Thus

$$\begin{aligned} M(r_n;\, f_1 + f_2) \geq \exp r_n^{\varrho_2-\varepsilon} - \exp r_n^{\varrho_1+\varepsilon} &= \exp r_n^{\varrho_2-\varepsilon}\{1 - \exp(r_n^{\varrho_1+\varepsilon} - r_n^{\varrho_2-\varepsilon})\} \\ &> \tfrac{1}{2}\exp r_n^{\varrho_2-\varepsilon} \end{aligned}$$

provided ε is so small that $\varrho_1 + \varepsilon < \varrho_2 - \varepsilon$ and n is sufficiently large. Thus $\varrho \geq \varrho_2$ and the order of the sum, $f_1 + f_2$, equals ϱ_2. □

NOTE. The proof also applies without significant changes when $\varrho_2 = \infty$. It should be noted that the theorem is sometimes false when $\varrho_1 = \varrho_2$.

Taking $f_1(z) = e^z$ and $f_2(z) = -e^z$, $\varrho_1 = \varrho_2 = 1$ and $\varrho = 0$. We can say that, if $\varrho_1 \leq \varrho_2$, then $\varrho \leq \varrho_2$.

4.2.2 Theorem. If $f_1(z)$ and $f_2(z)$ are entire functions of order ϱ_1 and ϱ_2, respectively, where $\varrho_1 \leq \varrho_2$, then the order ϱ of $F(z) = f_1(z)f_2(z)$ is such that $\varrho \leq \varrho_2$.

Proof. Given $\varepsilon > 0$ and r sufficiently large, we have

$$\begin{aligned} M(r; f_1 f_2) &\leq M(r; f_1)M(r; f_2) \\ &\leq \exp r^{\varrho_1+\varepsilon} \exp r^{\varrho_2+\varepsilon} \\ &\leq \exp 2r^{\varrho_2+\varepsilon} \end{aligned}$$

Thus $\varrho \leq \varrho_2 + \varepsilon$ and hence $\varrho \leq \varrho_2$. □

4.2.3 Theorem. If $f(z)$ is an entire function of order ϱ and $P(z)$ is a nonzero polynomial, then the product $f(z)P(z)$ is of order ϱ. If the quotient $f(z)/P(z)$ is an entire function, then it also is of order ϱ.

Proof. From the previous theorem the order of $f(z)P(z)$ does not exceed ϱ. Since $|P(z)| > 1$ for z sufficiently large, for these values of z, $|f(z)P(z)| \geq |f(z)|$. Thus the order of $f(z)P(z)$ is not less than ϱ, which proves the first part of the theorem. We note also that if $f(z)/P(z)$ is entire, then its order is the same as $P(z)f(z)/P(z) = f(z)$ and the second part of the theorem is proved. □

4.3 TYPE

Entire functions may be further subdivided as follows. Given an entire function of order ϱ (finite), suppose $\exists$ a positive k such that $M(r) < \exp kr^{\varrho}$ for $r > R_0$. Then $f(z)$ is said to be of finite type. The greatest lower bound $\sigma = \inf k \geq 0$ of the k's for which $M(r) < \exp kr^{\varrho}$ holds (starting from some sufficiently large r), is called the *type* of $f(z)$, e.g., e^z has order $\varrho = 1$ and type $\sigma = 1$.

4.3.1 Definition. $f(z)$ is called *normal, mean, or finite type* if $0 < \sigma < \infty$ and $f(z)$ is called *minimum type* if $\sigma = 0$. $f(z)$ is called *maximum type* or *infinite type* if $\sigma = \infty$ (i.e., $M(r)$ exceeds $\exp kr^{\varrho}$ for arbitrary large r).

As a consequence, we have the following result.

4.3.2 Theorem. The type σ of an entire function of order ϱ, $0 < \varrho < \infty$, is given by the formula

$$\sigma = \overline{\lim_{r\to\infty}} \frac{\log M(r)}{r^\varrho}.$$

Proof. Since $\sigma = \inf k$, given $\varepsilon > 0$, $\exists R(\varepsilon) > 0$ such that

$$M(r) < \exp(\sigma + \varepsilon)r^\varrho \qquad \text{for} \quad r > R(\varepsilon).$$

Also, there is a sequence $\{r_n\}$ such that

$$0 < r_1 < \cdots < r_n < \cdots \qquad \text{and} \quad M(r_n) > \exp(\sigma - \varepsilon)r_n{}^\varrho.$$

Thus

$$\log M(r)/r^\varrho < \sigma + \varepsilon \qquad \text{for} \quad r > R(\varepsilon)$$

while

$$\log M(r_n)/r_n{}^\varrho > \sigma - \varepsilon \qquad \text{for arbitrary large } r_n.$$

This is precisely what is meant by $\sigma = \overline{\lim}_{r\to\infty} \log M(r)/r^\varrho$. □

EXAMPLE.

$$(e^r - 1)/2 \leq M(r) = \max_{|z|=r} | \sin z | \leq (e^r + 1)/2$$

and $\sin z$ is of order $\varrho = 1$ and type $\sigma = 1$.

4.4 Note that the maximum modulus $M(r)$ describes the growth of an entire function in the neighborhood of the point at infinity but gives no information about the behavior of $f(z)$ in various unbounded subdomains. Consider for example e^z (order 1, type 1)

$$| e^z | = e^x = e^{r\cos\theta}.$$

In the closed angle $-(\pi/2) + \varepsilon \leq \theta \leq (\pi/2) - \varepsilon$, $\varepsilon > 0$,

$$e^{r\sin\varepsilon} \leq | e^z | \leq e^r$$

and thus $e^z \to \infty$ as $r \to \infty$. However, in every closed angle

$$(\pi/2) + \varepsilon \leq \theta \leq (3\pi/2) - \varepsilon, \qquad \varepsilon > 0,$$

$$| e^z | \leq e^{-r\sin\varepsilon}$$

and $e^z \to 0$ as $r \to \infty$. Thus $\exists$ two open angles each of π radians, viz., the right half-plane $\mathscr{R}(z) > 0$ and the left half-plane $\mathscr{R}(z) < 0$, such that $\lim_{r\to\infty} e^z = \infty$ in every angle in $\mathscr{R}(z) > 0$ while $\lim_{r\to\infty} e^z = 0$ in every angle in $\mathscr{R}(z) < 0$.

4.5 There are several types of "enumerative" functions, some of which will be studied in the section dealing with the so-called elementary Nevanlinna theory. The simplest of these is the function $n(r)$, viz., the number of zeros of $f(z)$ in $|z| \leq r$. We now give several theorems dealing with $n(r)$.

4.5.1 Theorem. If $f(z)$ is an entire function of order $\varrho < \infty$ and has an infinity of zeros with $f(0) \neq 0$, then given $\varepsilon > 0$, $\exists R_0$ such that for $R \geq R_0$,

$$n\left(\frac{R}{3}\right) \leq \frac{1}{\log 2} \cdot \log\left\{\frac{\exp R^{\varrho+\varepsilon}}{|f(0)|}\right\},$$

where $n(R)$ denotes the number of zeros of $f(z)$ in $|z| \leq R$.

Proof. If $f(z)$ is analytic in $|z| \leq R$ and $a_1, a_2, \ldots, a_n$ are zeros of $f(z)$ in $|z| \leq R/3$, then for

$$g(z) = \frac{f(z)}{\left(1 - \dfrac{z}{a_1}\right) \cdots \left(1 - \dfrac{z}{a_n}\right)},$$

we have $|g(z)| \leq M/2^n$, where $|f(z)| \leq M$ for $|z| = R$ and $|a_p| \leq R/3$, $p = 1, \ldots, n$. Thus (for $|z| = R$)

$$|z/a_p| \geq 3 \qquad \text{and} \qquad |1 - z/a_p| \geq 2.$$

By the maximum modulus theorem $|g(0)| \leq M/2^n$. Therefore $|f(0)| \leq M/2^n$ and

$$n \equiv n\left(\frac{R}{3}\right) \leq \frac{1}{\log 2} \cdot \log\left\{\frac{M}{|f(0)|}\right\}.$$

If, further, $f(z)$ is of order ϱ, then for $r > r_0$, $M(r) < \exp r^{\varrho+\varepsilon}$ and the result follows. □

4.5.2 Corollary.

I. $n(R) = O(R^{\varrho+\varepsilon})$. For

$$n\left(\frac{R}{3}\right) \leq \frac{1}{\log 2}\{R^{\varrho+\varepsilon} - \log|f(0)|\}$$

and

$$-\log |f(0)| < R_1^{\varrho+\varepsilon},$$

say, then for $R \geq R_1$

$$n\left(\frac{R}{3}\right) < \frac{2}{\log 2} R^{\varrho+\varepsilon}$$

and

$$n(R) = O(R^{\varrho+\varepsilon}).$$

II. Since $n(R)$ denotes the number of zeros for which $|a_n| \leq R$, then $n(R)$ is a nondecreasing function of R which is constant in intervals. It is zero for $R < |a_1|$ if $f(0)$ is not zero. By virtue of Jensen's formula,

$$\int_0^R \frac{n(x)}{x} dx = \frac{1}{2\pi} \int_0^{2\pi} \log |f(Re^{i\theta})| \, d\theta - \log |f(0)|.$$

Since $f(z)$ is of order ϱ, i.e., $|f(Re^{i\theta})| < k_1 \exp R^{\varrho+\varepsilon}$, $\varepsilon > 0$, then

$$\log |f(Re^{i\theta})| < kR^{\varrho+\varepsilon}.$$

Thus

$$\int_0^{2R} \frac{n(x)}{x} dx < k_2 R^{\varrho+\varepsilon}.$$

Since $n(R)$ is nondecreasing,

$$\int_0^{2R} \frac{n(x)}{x} dx \geq n(R) \int_R^{2R} \frac{dx}{x} = n(R) \log 2$$

and

$$n(R) \leq \frac{1}{\log 2} \int_0^{2R} \frac{n(x)}{x} dx < k_3 R^{\varrho+\varepsilon}.$$

Thus, roughly, the higher the order of a function the more zeros it may have in a given region.

4.6 Theorem. If $f(z)$ is of order $\varrho < \infty$ and $r_1, r_2, \ldots$ are the moduli of its infinite number of zeros, then $\sum_{(n)} 1/r_n^{\alpha} < \infty$ for $\alpha > \varrho$.

Proof. Let β be a number between α and ϱ, i.e., $\varrho < \beta < \alpha$. Then $n(r) < Ar^{\beta}$. Putting $r = r_n$, we have

$$n < Ar_n^{\beta} \qquad \text{and} \qquad r_n^{-\alpha} < A_1 n^{-\alpha/\beta}.$$

Since $\alpha/\beta > 1$,

$$\sum_{(n)} \frac{1}{r_n^{\alpha}} < \infty. \quad \square$$

NOTE. Clearly the result is trivial for a finite number of zeros.

4.7 Definition. The lower bound of the positive numbers α for which $\sum_{(n)} 1/r_n^{\alpha}$ is convergent, is called the *exponent of convergence of the zeros* and is denoted by ϱ_1. Formally, the empty set has $\varrho_1 = 0$ and if the series diverges for all positive α, then $\varrho_1 = \infty$. Thus

$$\sum_{(n)} \frac{1}{r_n^{\varrho_1+\varepsilon}} < \infty, \qquad \varepsilon > 0$$

and

$$\sum_{(n)} \frac{1}{r_n^{\varrho_1-\varepsilon}} = \infty, \qquad \varepsilon > 0.$$

We have proved that $\varrho_1 \leq \varrho$ since it may be possible to find numbers less than ϱ for which the series converges.

4.7.1 Lemma. The number defined by the equations

$$\varrho_1 = \overline{\lim_{n\to\infty}} \frac{\log n}{\log r_n} = \overline{\lim_{r\to\infty}} \frac{\log n(r)}{\log r}$$

has the property: if

$$\varrho_1 \text{ (finite)} = \overline{\lim_{n\to\infty}} \frac{\log n}{\log r_n} \qquad \left\{\text{or } \overline{\lim_{r\to\infty}} \frac{\log n(r)}{\log r}\right\},$$

then ϱ_1 is the exponent of convergence of the zeros of $f(z)$, i.e.,

$$\sum_{(n)} \frac{1}{r_n^{\varrho_1+\varepsilon}} < \infty, \qquad \varepsilon > 0$$

and

$$\sum_{(n)} \frac{1}{r_n^{\varrho_1-\varepsilon}} = \infty, \qquad \varepsilon > 0.$$

Proof. The limit implies that

$$\log n/\log r_n < \varrho_1 + \varepsilon \qquad \text{for} \quad n > N \tag{1}$$

and

$$\log n/\log r_n > \varrho_1 - \varepsilon \qquad \text{for} \quad n > N \tag{2}$$

From inequality (2), $n > r_n^{\varrho_1-\varepsilon}$ and thus $1/n < 1/r_n^{\varrho_1-\varepsilon}$ but since

$$\sum_{(n)} 1/n = \infty, \qquad \sum_{(n)} 1/r_n^{\varrho_1-\varepsilon} = \infty.$$

Using inequality (1), let $\varepsilon' > 0$ and let $\varepsilon = \frac{1}{2}\varepsilon'$, then $\exists N$ such that

$$n < r_n^{\varrho_1+\varepsilon}, \qquad \forall n > N.$$

Define

$$\delta \equiv \frac{\varepsilon' - \varepsilon}{\varrho_1 + \varepsilon} = \frac{\frac{1}{2}\varepsilon'}{\varrho_1 + \frac{1}{2}\varepsilon'}$$

and thus $\delta > 0$. Hence

$$\begin{aligned}
\varepsilon' &= \varepsilon + \varepsilon\delta + \varrho_1\delta \\
&\rightarrow \varrho_1 + \varepsilon' = (\varrho_1 + \varepsilon)(1 + \delta) \\
&\rightarrow r_n^{(\varrho_1+\varepsilon)(1+\delta)} > n^{1+\delta}, \qquad \forall n > N \\
&\rightarrow r_n^{\varrho_1+\varepsilon'} > n^{1+\delta}, \qquad \forall n > N \\
&\rightarrow \frac{1}{r_n^{\varrho_1+\varepsilon'}} < \frac{1}{n^{1+\delta}}, \qquad \forall n > N \\
&\rightarrow \sum_{(n)} \frac{1}{r_n^{\varrho_1+\varepsilon'}} < \infty. \quad \square
\end{aligned}$$

The series $\sum_{(n)} 1/r_n^{\varrho_1}$ may either converge or diverge. For example, take $r_n = n$ or $r_n = n(\log n)^2$. If the zeros of $f(z)$ are finite or nil, $\varrho_1 = 0$. Thus $\varrho_1 > 0$ implies $\exists$ infinitely many zeros.

NOTE. We can have $\varrho_1 < \varrho$, e.g., if $f(z) = e^z$, $\varrho = 1$ but since there are no zeros of e^z then $\varrho_1 = 0$. Let $f(z)$ be an entire function of order $\varrho < \infty$, $f(0) \neq 0$, $f(z_n) = 0$, $n = 1, 2, \ldots$. Then $\exists$ an integer $p + 1$ such that $\sum_{n=1}^{\infty} 1/|z_n|^{p+1} < \infty$. By the previous theorem any integer exceeding ϱ will serve as $p + 1$.

4.8 Definition. The smallest integer p for which $\sum_{n=1}^{\infty} 1/|z_n|^{p+1} < \infty$, is called the *genus* (rank) of the canonical product. The genus of the general entire function

$$f(z) = z^m e^{g(z)} \prod_{n=1}^{\infty} \left(1 - \frac{z}{z_n}\right) e^{Q_\nu(z/z_n)}$$

will be defined later. Sometimes the two will coincide. Thus by Weierstrass's theorem,

$$f(z) = e^{g(z)} z^m \prod_{n=1}^{\infty} \left(1 - \frac{z}{z_n}\right) \exp\left(\frac{z}{z_n} + \frac{z^2}{2z_n^2} + \cdots + \frac{z^\nu}{\nu \cdot z_n^\nu}\right)$$

where $\nu = p$. (We were previously looking for ν such that $\sum_{n=1}^{\infty} | z/z_n |^{\nu+1}$ would converge.) If the z_n's are finite, define $p = 0$ and the product as $\prod_{n=1}^{m}(1 - (z/z_n))$.

Examples. If $z_n = n$, then $p = 1$, $(\sum_{(n)} 1/| z_n |^2 < \infty)$; if $z_n = e^n$, then $p = 0$, and if $z_1 = \frac{1}{2} \log 2$, $z_n = \log n$, $n \geq 2 \ \exists$ no finite p.

4.8.1 Summarizing: For $\varrho_1 < \sigma$,

$$\sum_{n=1}^{\infty} 1/| z_n |^\sigma < \infty, \qquad \sum_{n=1}^{\infty} 1/| z_n |^{p+1} < \infty,$$

but

$$\sum_{n=1}^{\infty} 1/| z_n |^p = \infty$$

since p is the smallest integer for which the preceding equation holds. Thus, if ϱ_1 is not an integer, $p = [\varrho_1]$. If ϱ_1 is an integer, then either

$$\text{(i)} \quad \sum_{n=1}^{\infty} 1/| z_n |^{\varrho_1} < \infty$$

or

$$\text{(ii)} \quad \sum_{n=1}^{\infty} 1/| z_n |^{\varrho_1} = \infty.$$

4.8.2 Note that

$$\sum_{n=1}^{\infty} 1/| z_n |^{\varrho_1+\varepsilon} < \infty \qquad \text{for} \quad \varepsilon > 0$$

but we have no information for $\sum_{n=1}^{\infty} 1/| z_n |^{\varrho_1}$. It is useful to subdivide entire functions into two further classes depending upon whether the function $f(z)$ has zeros such that $\sum_{n=1}^{\infty} 1/r_n^{\varrho_1}$ converges or diverges, but we shall not pursue this subject of convergence or divergence class any further.

We have two cases to consider.

CASE I. $p + 1 = \varrho_1$,

CASE II. $p = \varrho_1$. Hence $p \leq \varrho_1$, but $\varrho_1 \leq \varrho$, thus

$$p \leq \varrho_1 \leq \varrho \qquad \text{and} \qquad p \leq \varrho_1 \leq p + 1$$

since $\sum_{n=1}^{\infty} 1/|z_n|^{p+1} < \infty$.

4.9 Hadamard's Factorization Theorem. If $f(z)$ is an entire function of order ϱ with zeros $z_1, z_2, \ldots$, $[f(0) \neq 0]$, then $f(z) = e^{Q(z)}P(z)$ where $P(z)$ is the canonical product formed with the zeros of $f(z)$ and $Q(z)$ is a polynomial of degree not greater than ϱ. (The canonical product of course includes the exponential convergence producing factor which may be unity.)

Proof. We have that since $f(z)$ is an entire function, $f(z) = f(0)P(z)e^{Q(z)}$, where $P(z)$ is a product of primary factors and $Q(z)$ is an entire function. We require to prove that $Q(z)$ is a polynomial. Let $\nu = [\varrho]$. Thus $p \leq \nu$. Taking logs and differentiating $\nu + 1$ times, we obtain

$$\frac{d^\nu}{dz^\nu}\left\{\frac{f'(z)}{f(z)}\right\} = Q^{(\nu+1)}(z) - \nu! \sum_{n=1}^{\infty} \frac{1}{(z_n - z)^{\nu+1}}.$$

{Note that

$$\frac{d^{\nu+1}}{dz^{\nu+1}} \sum_{(n)} \left(\frac{z}{z_n} + \cdots + \frac{1}{p}\left(\frac{z}{z_n}\right)^p\right) = 0\Big\}.$$

To show that $Q(z)$ is a polynomial of degree at most ν, we require to show that $Q^{(\nu+1)}(z) = 0$. Let

$$g_R(z) = \frac{f(z)}{f(0)} \prod_{|z_n| \leq R} \left(1 - \frac{z}{z_n}\right)^{-1}.$$

For $|z| = 2R$ and $|z_n| \leq R$, $|1 - (z/z_n)| \geq 1$ and thus

$$|g_R(z)| \leq |f(z)|/|f(0)| = O(\exp(2R)^{\varrho+\varepsilon}) \qquad \text{for} \quad |z| = 2R.$$

Since $g_R(z)$ is entire, $\{f(0) \neq 0, f(z)$ is entire and $\prod_{|z_n| \leq R}(1 - (z/z_n))^{-1}$ cancels with factors in $f(z)\}$, then $|g_R(z)| = O(\exp(2R)^{\varrho+\varepsilon})$ also for $|z| < 2R$ (by the maximum modulus theorem). Let $h_R(z) = \log g_R(z)$, the logarithms being determined for $h_R(0) = 0$. Then $h_R(z)$ is analytic for $|z| \leq R$ since $g_R(z) \neq 0$ in $|z| \leq R$ and $\mathscr{R}\{h_R(z)\} < KR^{\varrho+\varepsilon}$. (We have absorbed $2^{\varrho+\varepsilon}$ in K.) The real part may be negative but cannot be $-\infty$ in $|z| \leq R$.

By a previous result (Theorem 3.7.1),

$$| h_R^{(\nu+1)}(z) | \leq \frac{2^{\nu+3}(\nu+1)!R}{(R-r)^{\nu+2}} \cdot KR^{\varrho+\varepsilon} \qquad \text{for} \quad | z | = r < R$$

and we have

$$h_R^{(\nu+1)}(z) = O(R^{\varrho+\varepsilon-\nu-1}) \qquad \text{for} \quad | z | = \tfrac{1}{2}R = r$$

Hence

$$Q^{(\nu+1)}(z) = h_R^{(\nu+1)}(z) + \nu! \sum_{|z_n|>R} 1/(z_n - z)^{\nu+1}.$$

Since

$$h_R(z) = \log g_R(z) = \log f(z) - \log f(0) - \sum_{|z_n| \leq R} \log(1 - (z/z_n)),$$

then

$$h_R^{(\nu+1)}(z) = \frac{d^\nu}{dz^\nu} \left\{ \frac{f'(z)}{f(z)} \right\} + \nu! \sum_{|z_n| \leq R} \frac{1}{(z_n - z)^{\nu+1}}.$$

Thus

$$Q^{(\nu+1)}(z) = O(R^{\varrho+\varepsilon-\nu-1}) + O\Big(\sum_{|z_n|>R} | z_n |^{-\nu-1} \Big) \qquad \text{for} \quad | z | = \tfrac{1}{2}R$$

and so also for $| z | < \frac{1}{2} R$. Since $\nu = [\varrho]$, $\nu + 1 > \varrho$. Terms $O(R^{\varrho+\varepsilon-\nu-1}) \to 0$ as $R \to \infty$, provided ε is small. Also, since $\sum_{n=1}^{\infty} | z_n |^{-\nu-1}$ converges, terms $O(\sum_{|z_n|>R} | z_n |^{-\nu-1}) \to 0$ as $R \to \infty$ and $\sum_{|z_n|>R} | z_n |^{-\nu-1}$ becomes in effect the remainder term for R sufficiently large. Since $Q^{(\nu+1)}(z)$ is independent of R it must be zero, and the theorem follows. □

What we have shown is that $f(z) = e^{Q(z)}P(z)$ where $Q(z)$ is a polynomial of degree $\nu \leq \varrho$ and

$$P(z) = \prod_{n=1}^{\infty} \left(1 - \frac{z}{z_n}\right) \exp\!\left(\frac{z}{z_n} + \cdots + \frac{1}{p} \left(\frac{z}{z_n} \right)^p \right),$$

where p is the smallest integer for which $\sum_{n=1}^{\infty} 1/| z_n |^{p+1} < \infty$.

As an example of Weierstrass's theorem and Hadamard's theorem we express $\sin \pi z$ as an infinite product. The zeros are $z = \pm n$ and all are simple. Arrange the zeros in a sequence $0, +1, -1, +2, -2, \ldots$.

I. We consider

$$z \prod_{n=1}^{\infty} \left(1 - \frac{z}{n}\right) e^{z/n} \left(1 + \frac{z}{n}\right) e^{-z/n} \tag{1}$$

[the polynomial

$$Q_\nu\left(\frac{z}{z_n}\right) = \frac{z}{z_n} + \cdots + \frac{1}{\nu}\left(\frac{z}{z_n}\right)^\nu$$

has $\nu = p$ such that $\sum_{n=1}^\infty 1/|z_n|^{p+1}$ converges and p is integral, $p = 1$, and $\nu = 1$ and $Q_\nu(z/z_n) = z/n$]. Then (1) becomes $z\prod_{n=1}^\infty(1-(z^2/n^2))$ and

$$\sin \pi z = e^{g(z)} \cdot z \prod_{n=1}^\infty (1-(z^2/n^2)).$$

Taking the logarithmic derivative

$$\frac{\pi \cos \pi z}{\sin \pi z} = g'(z) + \frac{1}{z} + \sum_{n=1}^\infty \frac{2z}{z^2 - n^2}.$$

If we use the fact that

$$\pi \cot \pi z = \frac{1}{z} + \sum_{n=1}^\infty \frac{2z}{z^2 - n^2},$$

then $g'(z) = 0$ and $g(z) =$ constant. Since $\sin \pi z/\pi z \to 1$ as $z \to 0$, then

$$e^{g(z)} = \pi$$

and

$$\sin \pi z = \pi z \prod_{n=1}^\infty (1-(z^2/n^2)).$$

II. By Hadamard's theorem, since the order of $\sin \pi z$ is $\varrho = 1$, then $Q(z)$ is a polynomial of degree less than or equal to $\varrho = 1$ and hence $Q(z) = A + Bz$. Since

$$e^{Q(z)} = \frac{\sin \pi z}{z\prod_{n=1}^\infty (1-(z^2/n^2))}$$

$e^{Q(z)}$ is an even function and $e^{A+Bz} = e^{A-Bz}$, $e^{2Bz} = 1$ or $B = 0$. Taking the limit as $z \to 0$ we have $e^A = \pi$. Thus

$$e^{Q(z)} = \pi$$

and

$$\sin \pi z = \pi z \prod_{n=1}^\infty (1-(z^2/n^2)).$$

With Hadamard's theorem, we are now in a position to prove the following result.

4.9.1 Theorem. If $f(z)$ is an entire function of order ϱ and $g(z)$ is an entire function of order $\varrho' \leq \varrho$ and if the zeros of $g(z)$ are all zeros of $f(z)$, then $H(z) = f(z)/g(z)$ is of order ϱ, at most.

Proof. Writing $P_1(z)$, $P_2(z)$ to be the canonical product of $f(z)$ and $g(z)$, respectively, we have,

$$f(z) = P_1(z)e^{Q_1(z)} \quad \text{and} \quad g(z) = P_2(z)e^{Q_2(z)},$$

Q_1, Q_2 being appropriate polynomials. Thus

$$H(z) = P(z)e^{Q_1(z)-Q_2(z)}$$

where $P(z) = P_1(z)/P_2(z)$ is the canonical product formed from the zeros of $P_1(z)$ which are not zeros of $P_2(z)$. Since the exponent of convergence of a sequence is not increased by removing some of the terms, the exponent of convergence and hence the order of $P(z)$, does not exceed ϱ. Further, $Q_1(z) - Q_2(z)$ is a polynomial of degree not exceeding ϱ, thus the order of $H(z) = f(z)/g(z)$ is of order ϱ, at most. □

4.10 Theorem. The order of a canonical product equals the exponent of convergence of its zeros.

Proof. Since for any entire function $\varrho_1 \leq \varrho$, we require to prove that $\varrho \leq \varrho_1$ for a canonical product. Let the zeros be $z_1, z_2, \ldots,$ and k be a constant greater than 1. Let $P(z)$ be the canonical product, and we have

$$\begin{aligned}\log|P(z)| &= \sum_{|z_n|\leq kr} \log\left|\left(1-\frac{z}{z_n}\right)\exp\left(\frac{z}{z_n}+\cdots+\frac{1}{p}\left(\frac{z}{z_n}\right)^p\right)\right| \\ &\quad + \sum_{|z_n|>kr} \log\left|\left(1-\frac{z}{z_n}\right)\exp\left(\frac{z}{z_n}+\cdots+\frac{1}{p}\left(\frac{z}{z_n}\right)^p\right)\right| \\ &= \sum\nolimits_1 + \sum\nolimits_2.\end{aligned}$$

For $\sum_2$, since $|z| = r$ and $|z_n| > kr$, $|z/z_n| < 1$ and

$$\log\left(1-\frac{z}{z_n}\right)\exp\left(\frac{z}{z_n}+\cdots+\frac{1}{p}\left(\frac{z}{z_n}\right)^p\right) = -\frac{1}{p+1}\left(\frac{z}{z_n}\right)^{p+1} - \cdots.$$

Thus

$$\begin{aligned}&\left|\log\left(1-\frac{z}{z_n}\right)\exp\left(\frac{z}{z_n}+\cdots+\frac{1}{p}\left(\frac{z}{z_n}\right)^p\right)\right| \\ &\quad < \frac{1}{p+1}\left\{\left|\frac{z}{z_n}\right|^{p+1} + \left|\frac{z}{z_n}\right|^{p+2} + \cdots\right\} \\ &\quad = \frac{1}{p+1}\left\{\frac{|z/z_n|^{p+1}}{1-|z/z_n|}\right\} < K\left|\frac{z}{z_n}\right|^{p+1}.\end{aligned}$$

Also, $\log|f| = \mathscr{R}\{\log f\} \leq |\log f|$. Thus

$$\log\left|\left(1-\frac{z}{z_n}\right)\exp\left(\frac{z}{z_n}+\cdots+\frac{1}{p}\left(\frac{z}{z_n}\right)^p\right)\right| \leq A\left(\frac{z}{z_n}\right)^{p+1}$$

and

$$\sum_{|z_n|>kr}\log|\cdots| = \sum\nolimits_2 = O\left\{\sum_{|z_n|>kr}\left|\frac{z}{z_n}\right|^{p+1}\right\} = O\left\{|z|^{p+1}\sum_{|z_n|>kr}\frac{1}{|z_n|^{p+1}}\right\}.$$

If $p+1=\varrho_1$,

$$\sum\nolimits_2 = O\{|z|^{p+1}\} = O\{|z|^{\varrho_1}\} = O\{r^{\varrho_1}\},$$

(we recall that $p \leq \varrho_1 \leq p+1$). If $p+1 > \varrho_1+\varepsilon$ (recall that $\sum_{n=1}^{\infty} r_n^{-(\varrho_1+\varepsilon)}$ converges), and ε is small enough, then

$$|z|^{p+1}\sum_{|z_n|>kr}|z_n|^{-p-1} = |z|^{p+1}\sum_{|z_n|>kr}|z_n|^{\varrho_1+\varepsilon-p-1}|z_n|^{-\varrho_1-\varepsilon}$$

$$< |z|^{p+1}(kr)^{\varrho_1+\varepsilon-p-1}\sum_{|z_n|>kr}|z_n|^{-\varrho_1-\varepsilon} = O\{|z|^{\varrho_1+\varepsilon}\},$$

since

(i) $1/|z_n| < 1/kr$,

(ii) $r = |z|$

(iii) $\varrho_1+\varepsilon-p-1<0$.

In $\sum_1$, $|z/z_n| \geq 1/k$. Note that $|z/z_n|$ can be large but cannot be small. Since

$$\log\left|\left(1-\frac{z}{z_n}\right)\exp\left(\frac{z}{z_n}+\cdots+\frac{1}{p}\left(\frac{z}{z_n}\right)^p\right)\right|$$

$$\leq \log\left(1+\left|\frac{z}{z_n}\right|\right)+\left|\frac{z}{z_n}\right|+\cdots+\frac{1}{p}\left|\frac{z}{z_n}\right|^p$$

and

$$\log\left(1+\left|\frac{z}{z_n}\right|\right) < \left|\frac{z}{z_n}\right| \qquad \text{since} \quad 1+|x| < e^{|x|},$$

$$\log\left|\left(1-\frac{z}{z_n}\right)\exp\left(\frac{z}{z_n}+\cdots+\frac{1}{p}\left(\frac{z}{z_n}\right)^p\right)\right| < K\left|\frac{z}{z_n}\right|^p$$

where K depends on k only. Thus,

$$\sum_1 = O\left\{ \sum_{|z_n| \le kr} \left| \frac{z}{z_n} \right|^p \right\} = O\left\{ |z|^p \sum_{|z_n| \le kr} |z_n|^{\varrho_1+\varepsilon-p} |z_n|^{-\varrho_1-\varepsilon} \right\}$$

$$= O\left\{ |z|^p (kr)^{\varrho_1+\varepsilon-p} \sum_{|z_n| \le kr} |z_n|^{-\varrho_1-\varepsilon} \right\}$$

$$= O\{ |z|^{\varrho_1+\varepsilon} \},$$

since

(i) $|z_n|$ are bounded,

(ii) $\sum_{|z_n| \le kr} |z_n|^{-\varrho_1-\varepsilon}$ is a finite series,

(iii) $p \le \varrho_1, \quad p < \varrho_1 + \varepsilon$

(iv) $|z_n| \le k|z| = kr.$

Thus

$$\log |P(z)| = O\{ |z|^{\varrho_1+\varepsilon} \}$$

and

$$|P(z)| = O\{\exp r^{\varrho_1+\varepsilon}\}$$

from which we conclude that the order of $P(z)$, viz., ϱ is such that $\varrho \le \varrho_1$ and since $\varrho_1 \le \varrho$, $\varrho_1 = \varrho$. □

A particularly useful result is the following lemma.

4.10.1 Lemma. If ϱ is not an integer, $\varrho_1 = \varrho$.

Proof. In any case $\varrho_1 \le \varrho$. Suppose $\varrho_1 < \varrho$. Then $P(z)$ is of order ϱ_1, i.e., $P(z)$ is of order less than ϱ. If $Q(z)$ is of degree q, $e^{Q(z)}$ is of order $q \le \varrho$ but $q < \varrho$ since q is integral and ϱ is not. Hence $f(z)$ is the product of two functions each of order less than ϱ. Thus $f(z)$ is of order less than ϱ which contradicts the hypothesis that $f(z)$ is an entire function of order ϱ. □

A consequence is that a function of nonintegral order must have an infinity of zeros. (Since if the number of zeros is finite, $\varrho_1 = 0 = \varrho$.) Also, if the order is not integral, the function is dominated by $P(z)$, whereas if the order is integral, $P(z)$ may reduce to a polynomial or a constant and the order depends entirely on the factor $e^{Q(z)}$.

4.10.2 In any case, since $P(z)$ is of order ϱ_1 and $e^{Q(z)}$ is of order q, then

$$\varrho = \max\{q, \varrho_1\}.$$

4.11 Definition. The *genus* of the entire function $f(z)$ is the greater of the two integers p and q and is therefore an integer. Since $p \leq \varrho$ and $q \leq \varrho$, the genus does not exceed the order.

EXAMPLE 1. For the function

$$\sin z = z \prod_{n=1}^{\infty} \left(1 - \frac{z^2}{n^2\pi^2}\right),$$

(actually $\prod_{n=1}^{\infty} z(1 \pm (z/n\pi))e^{\pm z/n}$) the order of $e^{Q(z)}$ is $q = 0$, $e^{Q_\nu(z/z_n)} = e^{\pm z/n}$, and $p = 1$. The genus is $\max(p, q) = 1$. $\varrho_1 = 1$ since the series

$$\sum_{n=1}^{\infty} 1/(n\pi)^{1+\varepsilon} < \infty, \qquad \varepsilon > 0,$$

and the order $\varrho = \max(q, \varrho_1) = 1$. Hence the genus is 1 and the order is 1.

EXAMPLE 2. For the function

$$f(z) = \prod_{n=2}^{\infty} \left(1 - \frac{z}{n(\log n)^2}\right)$$

the order of $e^{Q(z)}$ is $q = 0$. The genus is $\max(p, q) = 0$ since $e^{Q_\nu(z/z_n)} = 1$. The order is $\max(0, 1) = 1$ since $\varrho_1 = 1$ for $\sum_{n=2}^{\infty} 1/n(\log n)^2 < \infty$. We need to establish that

$$\sum_{n=2}^{\infty} \frac{1}{\{n(\log n)^2\}^r} < \infty \qquad \text{for} \quad r \geq 1 \quad \text{only.}$$

(We can use for example, Gauss's test for infinite series.) Hence the genus is 0. And the order is 1.

4.12 If we have the power series representation of an entire function, we can calculate order and type fairly simply as shall be illustrated by the next two theorems. In order to study more sophisticated functions we will need Stirling's approximation for the gamma function. Both the approximation and the gamma function will be studied a little later.

4.12.1 Theorem. A necessary and sufficient condition that

$$f(z) = \sum_{n=0}^{\infty} a_n z^n$$

should be an entire function of order ϱ, is that

$$\lim_{n\to\infty} \left(\frac{\log(1/\mid a_n \mid)}{n \log n}\right) = \frac{1}{\varrho}.$$

Proof. We use the fact that $\sum_{n=0}^{\infty} | a_n z^n |$ does not differ much from its greatest term, and that $| f(z) |$ lies between the two. Let

$$\underline{\lim}_{n \to \infty} \left(\frac{\log(1/ | a_n |)}{n \log n} \right) = \mu$$

where μ is zero, positive, or ∞. Then for every $\varepsilon > 0$

$$\log \left(\frac{1}{| a_n |} \right) > (\mu - \varepsilon)\, n \log n \qquad \text{for} \quad n > n_0,$$

i.e., $| a_n | < n^{-n(\mu-\varepsilon)}$. If $\mu > 0$, $\sum_{n=0}^{\infty} a_n z^n$ converges for all z so that $f(z)$ is an entire function. If μ is finite,

$$| f(z) | < Ar^{n_0} + \sum_{n=n_0+1}^{\infty} r^n n^{-n(\mu-\varepsilon)} \qquad (r > 1).$$

Let $\sum_1$ denote the part of the last series for which $n \leq (2r)^{1/\mu-\varepsilon}$ and let $\sum_2$ be the remainder

$$\textstyle\sum_1 < \exp\{(2r)^{1/\mu-\varepsilon} \log r\} \sum_{(n)} n^{-n(\mu-\varepsilon)} < K \exp\{(2r)^{1/\mu-\varepsilon} \log r\}.$$

In $\sum_2$, $n > (2r)^{1/\mu-\varepsilon}$, thus

$$rn^{-(\mu-\varepsilon)} < r\{(2r)^{1/\mu-\varepsilon}\}^{-(\mu-\varepsilon)} = \tfrac{1}{2}$$

and $\sum_2 < \sum_{(n)} (\frac{1}{2})^n < 1$. Thus

$$| f(z) | < B \exp\{(2r)^{1/\mu-\varepsilon} \log r\}$$

and $\varrho \leq 1/(\mu - \varepsilon)$. Making $\varepsilon \to 0$, $\varrho \leq 1/\mu$. If $\mu = \infty$, the same argument with an arbitrarily large μ shows that $\varrho = 0$. On the other hand, given ε, $\exists$ a sequence of values of n for which

$$\log\left(\frac{1}{| a_n |} \right) < (\mu + \varepsilon)\, n \log n,$$

i.e.,

$$| a_n | > n^{-n(\mu+\varepsilon)} \qquad \text{or} \qquad | a_n |\, r^n > \{rn^{-(\mu+\varepsilon)}\}^n.$$

Take $r = (2n)^{\mu+\varepsilon}$ and solving for n,

$$| a_n | r^n > 2^{(\mu+\varepsilon)n} = \exp\{\tfrac{1}{2}(\mu + \varepsilon) r^{1/\mu+\varepsilon} \log 2\}.$$

Since Cauchy's inequality gives $M(r) \geq |a_n| r^n$, then for a sequence of values of r tending to ∞

$$M(r) > \exp\{Ar^{1/\mu+\varepsilon}\},$$

thus $\varrho \geq 1/(\mu + \varepsilon)$ and for $\varepsilon \to 0$, $\varrho \geq 1/\mu$, i.e., if $f(z)$ is an entire function, its order $\varrho = 1/\mu$ or

$$\varliminf_{n\to\infty} \left(\frac{\log(1/|a_n|)}{n \log n} \right) = \frac{1}{\varrho}. \quad \square$$

Further, if $\mu = 0$, then $f(z)$ is of infinite order. Let $f(z)$ be a function of finite order ϱ. Then $a_n \to 0$ and μ is nonegative and the argument has shown that $\mu = 1/\varrho$.

NOTE. If $f(z)$ is entire and if

$$\varliminf_{n\to\infty} \left(\frac{\log(1/|a_n|)}{n \log n} \right) = 0,$$

then $f(z)$ is of infinite order, since the limit as $n \to \infty$ can be 0 without $f(z)$ being entire.

A similar theorem for the type follows from the following lemma:

4.13 Lemma. Let $f(z)$ have a Taylor series expansion $\sum_{n=0}^{\infty} a_n z^n$. Suppose $\exists$ numbers $\mu > 0$, $\lambda > 0$ and an integer $N = N(\mu, \lambda) > 0$, such that $|a_n| < (e\mu\lambda/n)^{n/\mu}$ for all $n > N$. Then $f(z)$ is an entire function and given any $\varepsilon > 0$ there is a number $R = R(\varepsilon) > 0$ such that

$$M(r) < \exp\{(\lambda + \varepsilon)r^\mu\} \qquad \text{for all } r > R.$$

Proof. Since $|a_n| < (e\mu\lambda/n)^{n/\mu}$,

$$\sqrt[n]{|a_n|} < (e\mu\lambda/n)^{1/\mu} \qquad \text{for all } n > N.$$

Thus $\sqrt[n]{|a_n|} \to 0$, $n \to \infty$, and $f(z)$ is entire. Further,

$$\sqrt[n]{|a_n| r^n} < (e\mu\lambda/n)^{1/\mu}\, r < \tfrac{1}{2}$$

if $n > n_0 = n_0(r) = \{2^\mu e\mu\lambda r^\mu\}$. Choosing $R^1 = R^1(\mu, \lambda) > 1$ and so large that $n_0(r) > N$, if $r > R^1$, then

$$\sqrt[n]{|a_n| r^n} < \tfrac{1}{2} \qquad \text{or} \qquad |a_n| r^n < 1/2^n$$

provided $n > n_0$.

We now deduce an upper bound for $M(r)$.

$$\begin{aligned} M(r) = \max_{|z|=r} \left| \sum_{n=0}^{\infty} a_n z^n \right| &\leq \sum_{n=0}^{\infty} |a_n| r^n \\ &= \sum_{n=0}^{n_0} |a_n| r^n + \sum_{n=n_0+1}^{\infty} |a_n| r^n \\ &< \sum_{n=0}^{n_0} |a_n| r^n + \sum_{n=n_0+1}^{\infty} 1/2^n \\ &< \sum_{n=0}^{n_0} |a_n| r^n + 1 \qquad \text{if} \quad r > R^1. \end{aligned}$$

However,

$$\begin{aligned} \sum_{n=0}^{n_0} |a_n| r^n &= \sum_{n=0}^{N} |a_n| r^n + \sum_{n=N+1}^{n_0} |a_n| r^n \\ &< r^N \sum_{n=0}^{N} |a_n| + (n_0 - N) \max_{N+1 \leq n \leq n_0} |a_n| r^n, \end{aligned}$$

and

$$\begin{aligned} \max_{N+1 \leq n \leq n_0} |a_n| r^n \leq \max_{N+1 \leq n} |a_n| r^n &< \max_{N+1 \leq n} (e\mu\lambda/n)^{n/\mu} r^n \\ &\leq \max_{1 \leq n} (e\mu\lambda/n)^{n/\mu} r^n \\ &= \exp(\lambda r^\mu). \end{aligned}$$

The maximum is achieved for $n = \mu\lambda r^\mu$, thus

$$\max_{N+1 \leq n \leq n_0} |a_n| r^n < \exp(\lambda r^\mu).$$

Hence if $r > R^1$,

$$\begin{aligned} M(r) &< r^N \sum_{n=0}^{N} |a_n| + (n_0 - N) \exp(\lambda r^\mu) + 1 \\ &= r^N \sum_{n=0}^{N} |a_n| + (2^\mu e\mu\lambda r^\mu - N) \exp(\lambda r^\mu) + 1 \\ &= \exp(\lambda r^\mu)\left\{ 2^\mu e\mu\lambda r^\mu - N + r^N \exp(-\lambda r^\mu) \sum_{n=0}^{N} |a_n| + \exp(-\lambda r^\mu) \right\}. \end{aligned}$$

Given any $\varepsilon > 0$, $\exists$ a number $R = R(\varepsilon) > R^1$ such that the expression in brackets is less than $\exp \varepsilon r^\mu$ provided $r > R$. Hence

$$M(r) < \exp\{(\lambda + \varepsilon) r^\mu\} \qquad \text{for} \quad \text{all } r > R. \quad \square$$

4.13.1 Theorem. If $f(z)$ is an entire function of finite order ϱ $(0 < \varrho < \infty)$ and type σ, then

$$\sigma = \frac{1}{e\varrho} \overline{\lim_{n\to\infty}}\, n\,|a_n|^{\varrho/n}. \tag{1}$$

Proof. Suppose σ is finite. Then given any $k > \sigma$, $\exists$ a number $R = R(k) > 0$ such that $M(r) < \exp kr^{\varrho}$ for $r > R$. According to Cauchy's inequality,

$$|a_n| \leq \frac{M(r)}{r^n} < \frac{\exp(kr^{\varrho})}{r^n} \qquad \text{for all } r > R.$$

The minimum value of $\exp(kr^{\varrho})/r^n$ occurs for $r = (n/k\varrho)^{1/\varrho}$, thus

$$|a_n| < (e\varrho k/n)^{n/\varrho} \qquad \text{if } n > N \text{ and } r = (n/k\varrho)^{1/\varrho} > R(k).$$

Rewriting,

$$k > \frac{1}{e\varrho}\, n\,|a_n|^{\varrho/n}.$$

Therefore

$$k \geq \frac{1}{e\varrho} \overline{\lim_{n\to\infty}}\, n\,|a_n|^{\varrho/n}.$$

Since k is an arbitrary number exceeding σ,

$$\sigma \geq \frac{1}{e\varrho} \overline{\lim_{n\to\infty}}\, n\,|a_n|^{\varrho/n},$$

where the right-hand side is clearly finite. Now let k^1 be any number exceeding the right-hand side of (1). Then $\exists$ a number $N = N(k^1) > 0$ such that $|a_n| < (e\varrho k^1/n)^{n/\varrho}$ for all $n > N$. Applying the lemma with $\lambda = k^1$ and $\mu = \varrho$, given any ε, $\exists$ a number $R = R(\varepsilon) > 0$ [not to be confused with $R(\lambda)$], such that

$$M(r) < \exp\{(k^1 + \varepsilon)r^{\varrho}\} \qquad \text{for all } r > R.$$

Thus $\sigma \leq k^1$ and because of the choice of k^1,

$$\sigma \leq \frac{1}{e\varrho} \overline{\lim_{n\to\infty}}\, n\,|a_n|^{\varrho/n}.$$

Hence the result. Also, if the right-hand side of (1) is finite so is σ and if σ is infinite, so is the right-hand of (1). □

EXAMPLE 1. The function

$$f(z) = \sum_{n=1}^{\infty} (e\varrho\sigma/n)^{n/\varrho} z^n$$

is of order ϱ and type σ.

EXAMPLE 2. Since

$$\lim_{n\to\infty} \frac{\log n}{\log(1/\sqrt[n]{|a_n|})} = 0$$

characterizes an entire function of order zero, any function with coefficients $|a_n| = 1/n^{n/\varepsilon_n}$ where $\{\varepsilon_n\}$ is a sequence of positive numbers converging to zero is of order zero. For example,

$$f(z) = \sum_{n=1}^{\infty} \frac{z^n}{n^{n^{1+\delta}}}, \qquad \delta > 0$$

has $\varrho = 0$ (examine $\log(1/\sqrt[n]{|a_n|})$).

EXAMPLE 3. The condition

$$\overline{\lim_{n\to\infty}} \frac{\log n}{\log(1/\sqrt[n]{|a_n|})} = \infty$$

(same as

$$\underline{\lim}_{n\to\infty} \left(\frac{\log(1/|a_n|)}{n \log n} \right) = 0)$$

together with

$$\lim_{n\to\infty} \sqrt[n]{|a_n|} \to 0$$

characterizes an entire function of infinite order, e.g., consider $|a_n| = 1/n^{n\varepsilon_n}$ $\{\varepsilon_n\}$ a sequence of positive numbers converging to zero slowly enough that

$$\lim_{n\to\infty} \varepsilon_n \log n = \infty$$

(since we require $\lim_{n\to\infty} (1/\sqrt[n]{|a_n|}) \to \infty$).
The sequence $\varepsilon_n = 1/(\log n)^{1-\delta}$ $(n = 1, 2, \ldots)$ meets these requirements

if $0 < \delta < 1$ (since $\varepsilon_n \to 0$ but $\lim_{n\to\infty} \varepsilon_n \log n \to \infty$). Thus the series

$$f(z) = \sum_{n=0}^{\infty} \frac{z^n}{\exp(n^\delta \log n)}, \qquad 0 < \delta < 1$$

represents an entire function of infinite order.

4.14 We terminate the chapter with a paper of G. Pólya which is quoted almost verbatim. There is not a great deal which can be done to improve the explanation or substance. In order to prove the result to follow we need a theorem of H. Bohr† which says,

4.14.1 "Let ϱ be a number such that $0 < \varrho < 1$, and let $\omega = \phi(z)$ be any function analytic for $|z| \leq 1$ and satisfying

$$\phi(0) = 0, \qquad \max_{|z|=\varrho} |\phi(z)| = 1. \tag{1}$$

Let r_ϕ denote the radius of the largest circle $|\omega| = r_\phi$ whose points all represent values taken by $\phi(z)$ in the circular domain $|z| \leq 1$. Then r_ϕ is not less than C, $C = C(\varrho)$ being a positive number depending upon ϱ."

With this theorem, we can now prove the following (see Pólya‡).

4.14.2 Theorem. Suppose that $f(z)$, $g(z)$, $h(z)$ are entire functions connected by the relation

$$f(z) = g\{h(z)\}. \tag{2}$$

Suppose further that

$$h(0) = 0. \tag{3}$$

Let $F(r)$, $G(r)$, $H(r)$ denote the maximum moduli of $f(z)$, $g(z)$, $h(z)$, respectively, in the circle $|z| \leq r$. Then there is a definite number c, greater than 0 and less than 1, independent of $g(z)$, $h(z)$, and r, and such that

$$F(r) \geq G\{cH(\tfrac{1}{2}r)\}. \tag{4}$$

† "Über einen Satz von Edmund Landau," *Scripta Univ. Hierosolymitanarum* **1** No.2 (1923) 1–5.

‡ G. Pólya, "On an integral function of an integral function," *J. London Math. Soc.* **1** (1926), 12–15; J. E. Littlewood, "Lectures on the Theory of Functions," pp. 225–227. Oxford Univ. Press, London and New York, 1944.

We could substitute any positive fraction for $\frac{1}{2}$ provided c is replaced by some other suitable constant. The opposite inequality

$$F(r) \leq G\{H(r)\}$$

is an immediate consequence of the definition.

Proof. To fix our ideas let us take $\varrho = \frac{1}{2}$, put $C(\frac{1}{2}) = c$, and apply the theorem of Bohr to the function

$$\phi(z) = h(rz)/H(\tfrac{1}{2}r),$$

which satisfies the conditions (1). We see that the function $w = h(z)$ maps the circular domain $|z| \leq r$ on a Riemann surface extended over the w-plane whose various sheets cover the whole length of a certain circle of center $w = 0$ and of radius R, which is not less than $cH(\frac{1}{2}r)$.

Suppose that w_0 is a point on the circle $|w| = R$, such that

$$|g(w_0)| = G(|w_0|) = G(R).$$

Then there is at least one point z_0 inside $|z| \leq r$, such that

$$h(z_0) = w_0.$$

It follows that

$$G\{cH(\tfrac{1}{2}r)\} \leq G(|w_0|) = |g(w_0)| = |g\{h(z_0)\}| \leq F(r). \quad \square$$

We now prove the main result.

4.14.3 Theorem. If $g(z)$ and $h(z)$ are entire functions and $g\{(h(z)\}$ is an entire function of finite order, then there are only two possible cases: either (a) the internal function $h(z)$ is a polynomial and the external function $g(z)$ is of finite order, or else (b) the internal function $h(z)$ is not a polynomial but a function of finite order, and the external function $g(z)$ is of zero order.

Proof. The case where $g(z)$ or $h(z)$ is a constant is of no interest and will be excluded. Considering, if necessary, $h(z) - h(0)$ instead of $h(z)$, and $g\{w + h(0)\}$ instead of $g(w)$, we can and shall assume that (3) is true. Then we have, adopting the notation (2), the inequality (4). Observe that $F(r)$, $G(r)$, $H(r)$ are increasing functions.

We may express the hypothesis that $f(z)$ is of finite order by the inequality

$$F(r) < A \exp r^a \tag{5}$$

Put

$$h(z) = a_1 z + a_2 z^2 + \cdots + a_m z^m + \cdots,$$

and assume $|a_m| > 0$. We have

$$H(r) \geq |a_m| r^m \tag{6}$$

and by virtue of (4)–(6),

$$G(c\,|a_m|\,2^{-m}r^m) \leq G\{cH(\tfrac{1}{2}r)\} \leq F(r) < A \exp r^a$$
$$G(c\,|a_m|\,2^{-m}r) \leq A \exp r^{a/m}.$$

That is to say, the order of $g(z)$ does not exceed a/m. If $h(z)$ is not a polynomial, m can be chosen arbitrarily large and in this case the order of $g(z)$ is zero. In any case there is an inequality for $g(z)$, analogous to (6), let us say

$$G(r) \geq |b_n| r^n \qquad (|b_n| > 0, \quad n \geq 1).$$

Combining this with (4) and (5), we obtain

$$|b_n|\,c^n\{H(\tfrac{1}{2}r)\}^n \leq G\{cH(\tfrac{1}{2}r)\} \leq F(r) < A \exp r^a.$$

Thus the order of $h(z)$ is not greater than a. The chief point being settled by Theorem 4.14.2 is that there is naturally no difficulty in finding closer relations between the orders of magnitude of $F(r)$, $G(r)$, and $H(r)$. □

The case (b) of Theorem 4.14.3 is actually possible. Put

$$g(w) = 1 + 2^{-1}w + 2^{-4}w^2 + 2^{-9}w^3 + \cdots; \qquad h(z) = e^z.$$

The entire function

$$g\{h(z)\} = 1 + 2^{-1}e^z + 2^{-4}e^{2z} + \cdots \tag{7}$$

is the "upper half" of a theta series. The zeros and the order of magnitude of the whole theta series being perfectly known, we conclude on general principles that the function (7) is of the second order. We can easily obtain more precise information by direct calculation. Let $M(r)$ denote the maximum modulus and $n(r)$ the number of the zeros of the function (7) in the circle $|z| \leq r$. Then we have

$$\lim_{r\to\infty} r^{-2} \log M(r) = 2 \lim_{r\to\infty} r^{-2} n(r) = \frac{1}{4 \log 2}.$$

CHAPTER V

STANDARD FUNCTIONS AND CHARACTERIZATION THEOREMS

The gamma function is now studied in some detail, firstly with a view to consolidating some of the theorems we have developed and secondly, to illustrate the concept of analytic continuation. For further details, e.g., integrals involving $\Gamma(z)$, power series, relation to the zeta function, etc., the reader is referred to Whittaker and Watson.†

5.1 THE GAMMA FUNCTION AND ITS PROPERTIES

Define

$$\Gamma(x) = \int_{0+}^{\infty} e^{-t} t^{x-1}\, dt, \qquad x \text{ real}.$$

5.1.1 The integral converges at the upper limit since for all x,

$$t^{x-1}e^{-t} = t^{-2}t^{x+1}e^{-t} = O(t^{-2}), \qquad t \to \infty.$$

Then $\int_\delta^1 e^{-t}t^{x-1}\, dt$ does not converge for $x < 0$ if $0 < \delta < 1$ since for $t \in (0, 1)$,

$$e^t < e.$$

† E. T. Whittaker and G. N. Watson, "A Course of Modern Analysis." Cambridge Univ. Press, London and New York, 1962.

Thus $e^{-t} > e^{-1}$ and

$$\int_\delta^1 t^{x-1}e^{-t}\,dt > \int_\delta^1 t^{x-1}e^{-1}\,dt$$
$$= \frac{1}{e}\,\frac{t^x}{x}\Big]_\delta^1 = \frac{1}{xe}\{1-\delta^x\}$$

which diverges as $\delta \to 0$ for $x < 0$. Also, for $x > 0$, since $e^{-t} < 1$ for $t > 0$

$$\int_\delta^1 t^{x-1}e^{-t}\,dt < \int_\delta^1 t^{x-1}\,dt$$

which remains bounded.

5.1.2 The integral converges uniformly for $0 < a \leq x \leq b$, for

$$\int_0^\infty t^{x-1}e^{-t}\,dt = \int_0^1 + \int_1^\infty = O\left\{\int_0^1 t^{a-1}\,dt\right\} + O\left\{\int_1^\infty t^{b-1}e^{-t}\,dt\right\} = O(1)$$

independently of x. Hence the integral represents a continuous function for $x > 0$.

5.1.3 If z is complex, $\int_0^\infty t^{z-1}e^{-t}\,dt$ is again uniformly convergent over any finite region in which $\mathscr{R}(z) \geq a > 0$, for if $z = x + iy$

$$|\,t^{z-1}\,| = t^{x-1}$$

and we use 5.1.2. Hence $\Gamma(z)$ is analytic for $\mathscr{R}(z) > 0$.

5.1.4 For $x > 1$, integration by parts gives

$$\Gamma(x) = (x-1)\Gamma(x-1),$$
$$\Gamma(1) = \int_0^\infty e^{-t}\,dt = 1,$$

and

$$\Gamma(n) = (n-1)! \qquad \text{for positive integral } n.$$

5.1.5 $\Gamma(0^+) = +\infty$, for

$$\Gamma(x) > \int_{0^+}^1 t^{x-1}e^{-t}\,dt$$
$$> \frac{1}{e}\int_{0^+}^1 t^{x-1}\,dt = \frac{1}{ex} \to \infty, \qquad \text{as} \quad x \to 0^+.$$

Also

$$\lim_{x\to 0+} x\Gamma(x) = 1,$$

for $x\Gamma(x) = \Gamma(x+1)$, and since $\Gamma(x)$ is continuous,

$$\lim_{x\to 0+} \Gamma(x+1) = \Gamma(1) = 1.$$

5.1.6 For $x > 0$, $y > 0$,

$$\frac{\Gamma(x)\Gamma(y)}{\Gamma(x+y)} = \int_0^\infty \frac{t^{y-1}}{(1+t)^{x+y}}\,dt.$$

Since

$$\Gamma(x)\Gamma(y) = \int_0^\infty t^{x-1}e^{-t}\,dt \int_0^\infty s^{y-1}e^{-s}\,ds \qquad \text{for} \quad x > 0, \quad y > 0,$$

put $s = tv$ and

$$\Gamma(x)\Gamma(y) = \int_0^\infty t^{x-1}e^{-t}\,dt \int_0^\infty t^y v^{y-1}e^{-tv}\,dv$$

$$= \int_0^\infty v^{y-1}\,dv \int_0^\infty t^{x+y-1}e^{-t(1+v)}\,dt.$$

Letting $u = t(1+v)$

$$\Gamma(x)\Gamma(y) = \int_0^\infty v^{y-1}\,dv \int_0^\infty u^{x+y-1}e^{-u}(1+v)^{-x-y}\,du$$

$$= \Gamma(x+y)\int_0^\infty \frac{v^{y-1}}{(1+v)^{x+y}}\,dv.$$

Inversion of the integrals is justified, since the individual integrals converge uniformly for $x \geq \varepsilon > 0$ and $y \geq \varepsilon > 0$. Sometimes $\Gamma(x)\Gamma(y)/\Gamma(x+y)$ is called the *Beta function* $B(x, y)$ which equals $\int_0^1 t^{x-1}(1-t)^{y-1}\,dt$ by a suitable transformation.

5.1.7 Putting $x = y = \frac{1}{2}$, $v = \tan^2\theta$, we obtain

$$\{\Gamma(\tfrac{1}{2})\}^2 = 2\int_0^{\pi/2} d\theta = \pi.$$

Since $\Gamma(\frac{1}{2}) > 0$, we obtain $\Gamma(\frac{1}{2}) = \sqrt{\pi}$. Also, putting $y = 1 - x$,

$$\Gamma(x)\Gamma(1-x) = \int_0^\infty \frac{u^{-x}}{1+u}\,du = \frac{\pi}{\sin(1-x)\pi}, \qquad 0 < x < 1$$

and

$$\int_0^\infty \frac{x^{a-1}}{1+x}\,dx = \frac{\pi}{\sin a\pi} \qquad \text{for} \quad 0 < a < 1$$

by contour integration. Thus

$$\Gamma(x)\Gamma(1-x) = \frac{\pi}{\sin \pi x} \qquad \text{for} \quad 0 < x < 1.$$

5.1.8 Asymptotic Behavior of $\Gamma(x)$: Stirling's Formula. Consider $\Gamma(x)$, where x is an integer, say n, then $\Gamma(n) = (n-1)!$. We have

$$\log(n-1)! = \sum_{\nu=1}^{n-1} \log \nu,$$

also

$$\begin{aligned} \int_{\nu-1/2}^{\nu+1/2} \log t\,dt &= \int_0^{1/2} \{\log(\nu+t) + \log(\nu-t)\}\,dt \\ &= \int_0^{1/2} \{\log \nu^2 + \log(1 - t^2/\nu^2)\}\,dt \\ &= \log \nu + C_\nu, \qquad C_\nu = O(1/\nu^2). \end{aligned}$$

Thus

$$\begin{aligned} \log \Gamma(n) &= \log(n-1)! \\ &= \int_{1/2}^{n-1/2} \log t\,dt - \sum_{\nu=1}^{n-1} C_\nu \\ &= (n-\tfrac{1}{2})\log(n-\tfrac{1}{2}) - (n-\tfrac{1}{2}) - \tfrac{1}{2}\log\tfrac{1}{2} + \tfrac{1}{2} - \sum_{\nu=1}^{\infty} C_\nu + o(1) \\ &= (n-\tfrac{1}{2})\log n - n + C + o(1), \end{aligned}$$

where C is a constant.

Before establishing the nonintegral case, we prove the following lemma.

5.1.8.1 Lemma. For n large

$$\Gamma(n)/\Gamma(n+a) \approx n^{-a}.$$

Proof.

$$\lim_{n\to\infty} n^a\, \Gamma(a)\Gamma(n)/\Gamma(a+n) = \lim_{n\to\infty} n^a \int_0^1 t^{a-1}(1-t)^{n-1}\,dt.$$

Transform $t = v/n$. Then the right-hand side is

$$\lim_{n\to\infty} \int_0^n v^{a-1}[1-(v/n)]^{n-1}\,dv = \int_0^\infty v^{a-1}e^{-v}\,dv = \Gamma(a).$$

Thus

$$\lim_{n\to\infty} n^a\Gamma(n)/\Gamma(a+n) = 1$$

and $\Gamma(n)/\Gamma(n+a) \approx n^{-a}$ for large n. □

If x is not an integer, let $x = n + a$ where n is integral and $0 < a < 1$. From the previous lemma we have that since $\Gamma(n+a) \approx \Gamma(n)n^a$,

$$\begin{aligned}\log\Gamma(x) = \log\Gamma(n+a) &= \log\Gamma(n) + a\log n + o(1)\\ &= (n-\tfrac{1}{2})\log n - n + C' + a\log n + o(1)\\ &= (x-a-\tfrac{1}{2})\log(x-a) - x + a + C'\\ &\quad + a\log(x-a) + o(1)\\ &= (x-\tfrac{1}{2})\log x - x + C + o(1).\end{aligned}$$

To evaluate C, consider $\Gamma(2x)\Gamma(\frac{1}{2}) = 2^{2x-1}\Gamma(x)\Gamma(x+\frac{1}{2})$, a recurrence relation obtained by considering, for example, $B(x, x)$. Taking logs,

$$\begin{aligned}&(2x-\tfrac{1}{2})\log 2x - 2x + C + o(1) + \log\sqrt{\pi}\\ &\quad = (2x-1)\log 2 + (x-\tfrac{1}{2})\log x - x + C + o(1) + x\log(x+\tfrac{1}{2})\\ &\quad\quad - x - \tfrac{1}{2} + C + o(1).\end{aligned}$$

Thus

$$\begin{aligned}&2x\log 2 + 2x\log x - \tfrac{1}{2}\log 2x - 2x + C + o(1) + \log\sqrt{\pi}\\ &\quad = 2x\log 2 - \log 2 + x\log x - \tfrac{1}{2}\log x\\ &\quad\quad - 2x + 2C - \tfrac{1}{2} + x\log x\cdot\left(1+\frac{1}{2x}\right) + o(1)\end{aligned}$$

and

$$\log\sqrt{\pi} + \tfrac{1}{2}\log 2 = C - \tfrac{1}{2} + x\cdot\frac{1}{2x} + o(1) = C + o(1),$$

giving $C = \log\sqrt{2\pi} + o(1)$. Hence

$$\log\Gamma(x) = (x-\tfrac{1}{2})\log x - x + \log\sqrt{2\pi} + o(1)$$

and for x not a negative integer, since $e^{o(1)} = 1 + o(1)$,

$$\Gamma(x) \approx x^{x-1/2}e^{-x}\sqrt{2\pi}\,[1 + o(1)].$$

Stirling's formula for complex z can be shown,† the result holding uniformly for $-\pi + \delta \leq \arg z \leq \pi + \delta$, $\delta > 0$. The negative real axis is excluded since $\Gamma(z)$ has an infinite number of poles on it.

5.2 ANALYTIC CONTINUATION OF Γ(z)

For $\mathscr{R}(z) > 0$, $\Gamma(z) = \int_0^\infty e^{-t} t^{z-1}\, dt$ is an analytic function. We require now to extend analytically into the rest of the complex z-plane. Consider

$$I(z) \equiv \int_C e^{-\xi}(-\xi)^{z-1}\, d\xi,$$

where C is the real axis from ∞ to $\delta > 0$, $|\xi| = \delta$ in the positive sense, and the real axis from δ to ∞ (Fig. 1). Define $(-\xi)^{z-1} = e^{(z-1)\log(-\xi)}$ and choose

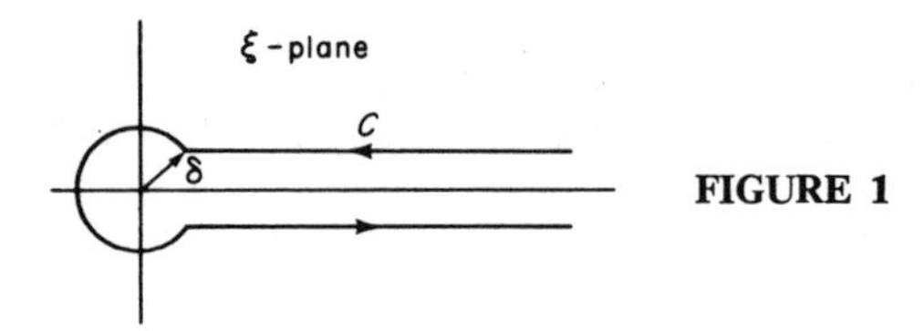

FIGURE 1

$\log(-\xi)$ to be real for $\xi = -\delta$. The integral converges uniformly in any finite region of the z-plane (since the integral depends upon C and the circle does not pass through the origin, i.e., δ is fixed.) Thus $I(z)$ is analytic for all finite values of z.

To evaluate $I(z)$, set $\xi = \varrho e^{i\phi}$. Then

$$\log(-\xi) = \log \varrho + i(\phi - \pi)$$

on the contour (so as to make $\log(-\xi)$ real for $\xi = -\delta$). The inetgrals on the portions of C, consisting of (∞, δ) and (δ, ∞) give

$$\int_\infty^\delta e^{-\varrho+(z-1)\{\log \varrho - i\pi\}}\, d\varrho + \int_\delta^\infty e^{-\varrho+(z-1)\{\log \varrho + i\pi\}}\, d\varrho$$

$$= \int_\delta^\infty e^{-\varrho+(z-1)\log \varrho}\{e^{(z-1)i\pi} - e^{-(z-1)i\pi}\}\, d\varrho$$

$$= -2i \sin \pi z \int_\delta^\infty e^{-\varrho} \varrho^{z-1}\, d\varrho.$$

† See E. C. Titchmarsh, "The Theory of Functions." Oxford Univ. Press, London and New York, 1939.

Also, on the circle $|\xi| = \delta$,

$$
\begin{aligned}
|(-\xi)^{z-1}| = |e^{(z-1)\log(-\xi)}| &= |e^{(z-1)[\log\delta + i(\phi-\pi)]}| \\
&= e^{(x-1)\log\delta - y(\phi-\pi)} \\
&= O(\delta^{x-1}),
\end{aligned}
$$

and the integral around the circle $|\xi| = \delta$ gives

$$O(\delta^x) = o(1) \qquad \text{as} \quad \delta \to 0$$

if $x > 0$ $(\leq \delta^{x-1} \int_0^{2\pi} d\phi)$. Letting $\delta \to 0$ we obtain

$$
\begin{aligned}
I(z) &= -2i \sin \pi z \int_0^\infty e^{-\varrho} \varrho^{z-1}\, d\varrho, \qquad \mathscr{R}(z) > 0 \\
&= -2i \sin \pi z\, \Gamma(z).
\end{aligned}
$$

Since $I(z)$ is analytic for all finite z, $\frac{1}{2} i I(z) \csc \pi z$ is analytic except possibly for poles of $\csc \pi z$. Further,

$$\tfrac{1}{2} i I(z) \csc \pi z = \Gamma(z), \qquad \mathscr{R}(z) > 0.$$

Hence $\frac{1}{2} i I(z) \csc \pi z$ is the analytic continuation of $\Gamma(z)$ in the entire z-plane.

Since the poles of $\csc \pi z$ are $z = 0, \pm 1, \pm 2, \ldots$ and $\Gamma(z)$ is analytic at $z = 1, 2, \ldots$, the only possible poles of $\frac{1}{2} i I(z) \csc \pi z$ are $z = 0, -1, -2, \ldots$. These are actually poles of $\Gamma(z)$, for if z is one of these numbers, say $-n$, then $(-\xi)^{z-1}$ is single-valued in C and $I(z)$ can be evaluated directly by Cauchy's theorem

$$(-1)^{n+1} \int_C \frac{e^{-\xi}}{\xi^{n+1}}\, d\xi = \frac{2\pi i}{n!} (-1)^{n+n+1} = \frac{-2\pi i}{n!}.$$

Thus $I(-n) = -2\pi i/n!$ and the poles of $\csc \pi z$ at $z = 0, -n$ are actually poles of $\Gamma(z)$. $(\Gamma(-n) = I(-n) \cdot 1/(-2i \sin n\pi).)$ The residue at $z = -n$ is

$$\lim_{z \to -n} \left\{ \left(\frac{-2\pi i}{n!} \right) \frac{z+n}{-2i \sin \pi z} \right\} = \frac{(-1)^n}{n!},$$

(actually

$$\left. \lim_{z \to -n} I(z) \frac{i}{2} \csc \pi z \cdot (z+n) = I(-n) \frac{i}{2} \lim_{z \to -n} \frac{z+n}{\sin \pi z} \right).$$

The formula $\Gamma(x)\Gamma(1-x) = \pi \csc \pi x$ and others, can now be justified for complex values. Thus $\Gamma(z)\Gamma(1-z) = \pi \csc \pi z$ for all nonintegral z (since the left-hand side is $\pi \csc \pi z$ for $0 < z < 1$ and $\Gamma(z)$ has an analytic continuation to the whole plane). Thus $1/\Gamma(z)$ is an entire function (since poles of $\Gamma(1-z)$ are canceled by zeros of $\sin \pi z$ or

$$\frac{1}{\Gamma(z)} = \frac{1}{\pi} \sin \pi z \frac{1}{2} i \frac{I(1-z)}{\sin \pi(1-z)} = \frac{i}{2\pi} I(1-z)$$

and $I(1-z)$ is analytic everywhere in the finite plane).

5.2.1 To prove $1/\Gamma(z)$ is of order 1.

Proof. Since $1/\Gamma(z) = I(1-z)/(-2i\pi)$, consider

$$I(z) = \int_C (-\omega)^{z-1} e^{-\omega} \, d\omega,$$

i.e.,

$$I(1-z) = \int_C (-\omega)^{-z} e^{-\omega} \, d\omega .$$

Take C the unit circle together with the real axis from 1 to ∞ described twice. On the circle, $-\omega = e^{i\phi}$, $-\pi \leq \phi \leq \pi$ and

$$|(-\omega)^{-z}| = |\exp(-i\phi re^{i\phi})| \leq e^{\pi r}.$$

The integral around the circle is $O(e^{\pi r})$, $\{|e^{-\omega}| \leq e\}$. On the rest of the contour, $-\omega = te^{i\phi}$, $t > 1$, $\phi = \pm\pi$. Therefore for this part of the integral

$$\left|\int_1^\infty (-\omega)^{-z} e^{-\omega} \, d\omega\right| \leq \int_1^\infty t^{-r} e^{-t} e^{\pi r} \, dt$$
$$= O\left\{e^{\pi r} \int_1^\infty e^{-t} t^r \, dt\right\}.$$

If $n = [r]$,

$$\int_1^\infty e^{-t} t^r \, dt < \int_0^\infty e^{-t} t^{n+1} \, dt = (n+1)! < (n+1)^{n+1} < (r+2)^{r+2}$$

(since $n < r + 1$), and $(r+2)^{r+2} = O(\exp r^{1+\varepsilon})$ by taking logs and observing that terms $O(r \log r) = O(r^{1+\varepsilon})$, $\varepsilon > 0$. Also, all terms including $O(e^{\pi r})$ are $O(\exp r^{1+\varepsilon})$, thus $\varrho \leq 1$. Since $\varrho \geq \varrho_1 = 1$ [observing the poles of $\Gamma(z)$] we conclude that $\varrho = 1$. ☐

We are now in a position to develop an expansion for $1/\Gamma(z)$, viz.

$$\frac{1}{\Gamma(z)} = e^{az+b}\, z \prod_{n=1}^{\infty} \left(1 + \frac{z}{n}\right) e^{-z/n}.$$

Since $\Gamma(1) = 1$,

$$\lim_{z\to 0} \frac{1}{z\Gamma(z)} = \lim_{z\to 0} \frac{1}{\Gamma(z+1)} = 1$$

and $1 = e^b$, $b = 0$. Putting $z = 1$,

$$1 = e^a \prod_{n=1}^{\infty} \left(1 + \frac{1}{n}\right) e^{-1/n}$$

Taking logs,

$$0 = a + \log \prod_{n=1}^{\infty} \left(1 + \frac{1}{n}\right) e^{-1/n}$$

and

$$a = -\left\{(\log 2 - 1) + \left(\log \frac{3}{2} - \frac{1}{2}\right) + \cdots\right\}$$
$$= \lim_{n\to\infty} \left\{1 + \frac{1}{2} \cdots + \frac{1}{n} - \log n\right\} = \gamma.$$

(For the existence of the limit we can use, for example, the integral test for series.) Thus

$$\frac{1}{\Gamma(z)} = e^{\gamma z} z \prod_{n=1}^{\infty} \left(1 + \frac{z}{n}\right) e^{-z/n}$$

Further, since

$$\gamma = \lim_{n\to\infty} \left(\sum_{\nu=1}^{n} \frac{1}{\nu} - \log n\right),$$

$$\frac{1}{\Gamma(z)} = \lim_{n\to\infty} \left\{\exp\left[\left(\sum_{\nu=1}^{n} \frac{1}{\nu} - \log n\right) z\right] z \prod_{\nu=1}^{n} \left(\frac{z+\nu}{\nu}\right) \exp\left(-z \sum_{\nu=1}^{n} \frac{1}{\nu}\right)\right\}.$$

5.2.2 Also, according to Gauss,

$$\Gamma(z) = \lim_{n\to\infty} \frac{n^z n!}{z(z+1)(z+2)\cdots(z+n)}.$$

It follows easily now that $\Gamma(z+1) = z\Gamma(z)$. Similar formulas can be easily established.

5.3 CONJUGATE POINTS

The gamma function assumes conjugate values at conjugate points, a property which facilitates calculation of absolute values. Thus, for example,

$$|\Gamma(\tfrac{1}{2} + it)|^2 = \Gamma(\tfrac{1}{2} + it)\Gamma(\tfrac{1}{2} - it)$$

and since

$$\Gamma(s)\Gamma(1 - s) = \pi/\sin \pi s$$

$$|\Gamma(\tfrac{1}{2} + it)|^2 = \pi/\sin \pi(\tfrac{1}{2} + it) = \pi/\cos i\pi t$$

and

$$|\Gamma(\tfrac{1}{2} + it)| = \sqrt{2\pi/(e^{\pi t} + e^{-\pi t})}\,.$$

We can similarly establish the asymptotic behavior of $|\Gamma(x + iy)|$ for large y, showing that

$$|\Gamma(x + iy)| \sim \sqrt{2\pi}\,|y|^{x-1/2}e^{-\pi|y|/2}$$

for finite x.

One or two examples now follow which illustrate, in particular, the use of previous theorems and lemmas in the calculation of order.

For a particularly practical function, we consider the following.

5.4 BESSEL'S FUNCTION $J_\nu(z)$

For ν a nonnegative integer

$$J_\nu(z) = \sum_{n=0}^{\infty} \frac{(-1)^n (z/2)^{\nu+2n}}{n!\,\Gamma(\nu + n + 1)}.$$

Thus $z^{-\nu}J_\nu(z)$ is an entire function since the series converges absolutely for all z. To calculate the order:

$$a_n = \begin{cases} 0 & \text{for } n \text{ odd} \\ \dfrac{(-1)^{n/2}}{2^n (n/2)!\,\Gamma(\nu + (n/2) + 1)} & \text{for } n \text{ even}, \end{cases}$$

thus

$$\frac{1}{\varrho} = \varliminf_{n\to\infty} \left(\frac{\log(1/|a_n|)}{n \log n}\right)$$

$$= \varliminf_{n\to\infty} \left(\frac{n \log 2 + \log(n/2)! + \log \Gamma(\nu + (n/2) + 1)}{n \log n}\right).$$

Since

$$\log \Gamma(n) = (n - \tfrac{1}{2}) \log n - n + C + o(1) \qquad \text{for} \quad C \text{ constant},$$

$$\frac{1}{\varrho} = \lim_{n\to\infty} \left(\frac{n \log 2 + \frac{1}{2}n \log n + O(n) + \frac{1}{2}n \log n + O(n)}{n \log n} \right) = 1 \quad \text{and} \quad \varrho = 1.$$

Hence $z^{-\nu/2}J_\nu(\sqrt{z})$ is an entire function of order $\frac{1}{2}$ [for

$$z^{-\nu}J_\nu(z) = O(\exp |z|^{1+\varepsilon}),$$

changing $z \to \sqrt{z}$ gives the result]. Thus $z^{-\nu/2}J_\nu(\sqrt{z})$ has an infinity of zeros and further, $J_\nu(z)$ has an infinity of zeros (for if $J_\nu(\sqrt{z}) = 0$ for $z = a$, $J_\nu(z) = 0$ for $z = \sqrt{a}$).

5.5 The following example requires a little more care. We require to calculate the order of

$$F_\alpha(z) = \int_0^\infty \exp(-t^\alpha) \cos zt\, dt \qquad (\alpha > 1).$$

For $\alpha > 1$ the integral converges uniformly for all finite z, thus $F_\alpha(z)$ is an entire function. Also,

$$\begin{aligned} F_\alpha(z) &= \int_0^\infty \exp(-t^\alpha) \left\{ \sum_{n=0}^\infty \frac{(-1)^n z^{2n} t^{2n}}{(2n)!} \right\} dt \\ &= \sum_{n=0}^\infty \frac{(-1)^n z^{2n}}{(2n)!} \int_0^\infty \exp(-t^\alpha) t^{2n}\, dt, \end{aligned}$$

where inversion is justified by convergence of

$$\int_0^\infty \exp(-t^\alpha) \left(\sum_{n=0}^\infty \frac{r^{2n} t^{2n}}{(2n)!} \right) dt = \int_0^\infty \exp(-t^\alpha) \cosh rt\, dt.$$

Now

$$\int_0^\infty \exp(-t^\alpha) t^{2n}\, dt = \frac{1}{\alpha} \int_0^\infty e^{-u} u^{(2n+1/\alpha)-1} du = \frac{1}{\alpha}\, \Gamma\left(\frac{2n+1}{\alpha}\right).$$

Thus

$$F_\alpha(z) = \frac{1}{\alpha} \sum_{n=0}^\infty \frac{(-1)^n}{(2n)!}\, \Gamma\left(\frac{2n+1}{\alpha}\right) z^{2n},$$

and $a_n = 0$, n odd. For n even,

$$\log \frac{1}{|a_n|} = \log n! - \log \Gamma\left\{\frac{n+1}{\alpha}\right\} + \log \alpha$$

$$= O(n) + n \log n - \left(\frac{n+1}{\alpha}\right) \log \left(\frac{n+1}{\alpha}\right) + O(n)$$

$$= n \log n - \frac{n}{\alpha} \log n + O(n)$$

and

$$\frac{1}{\varrho} = \lim_{n\to\infty} \left(\frac{\log (1/|a_n|)}{n \log n}\right) = 1 - \frac{1}{\alpha}.$$

Hence $\varrho = \alpha/(\alpha - 1)$. For $\alpha = 2$,

$$F_2(z) = \tfrac{1}{2}\sqrt{\pi} \exp(-z^2/4),$$

showing directly that $\varrho = 2$.

5.5.1 The convergence of this integral is not particularly obvious and we study it a little further. If $w = u + iv$,

$$|e^{iw}| = e^{-v} \qquad \text{and} \qquad |e^{-iw}| = e^{v}.$$

Thus if v is large and positive,

$$\cos w \approx \tfrac{1}{2} e^{-iw} \qquad \text{and} \qquad |\cos w| \approx \tfrac{1}{2} e^{v}.$$

If v is large and negative,

$$|\cos w| \approx \tfrac{1}{2} e^{-v}.$$

In the integral

$$F_\alpha(z) = \int_0^\infty \exp(-t^\alpha) \cos zt \, dt,$$

the factor $\cos zt$ may be large in absolute value, in fact it *will* be large unless z is (exactly) real. For $z = x + iy$, $y > 0$, t real,

$$\cos zt \approx \tfrac{1}{2} e^{-izt} = \tfrac{1}{2} e^{ty}(e^{-itx})$$

and

$$|\cos zt| \approx \tfrac{1}{2} e^{ty} \leq \tfrac{1}{2} e^{t|z|}.$$

However $|\cos zt|$ is no bigger than this, thus except for a factor $\frac{1}{2}$, $|\cos zt| \leq e^{t|z|}$ and for $\alpha > 1$,

$$|\exp(-t^{\alpha}) \cos zt| \leq \exp(t|z| - t^{\alpha}) \leq \exp(-t)$$

for all t for which $t^{\alpha-1} > 1 + |z|$. Hence the integral is uniformly absolutely convergent in any bounded range of z. For $\alpha = 1$, $|z| < 1$,

$$F_{\alpha}(z) = \frac{1}{2}\left\{\frac{1}{1 - iz} + \frac{1}{1 + iz}\right\}$$

tending to poles at $z = \pm i$, thus the integral converges only between the lines $y = \pm i$.

We now consider the derivative $f'(z)$ of an entire function and the following theorem is not unexpected.

5.6 Theorem. The derived entire function $f'(z)$ is of the same order (and type) as $f(z)$.

Proof. Let $M'(r) = \max_{|z|=r} f'(z)$. Then

$$\frac{M(r) - f(0)}{r} \leq M'(r) \leq \frac{M(R)}{R - r}$$

since

$$f(z) = \int_0^z f'(t)\, dt + f(0)$$

and by taking the integral along the straight line we have

$$M(r) \leq \int_0^{|z|} |f'(t)|\, dt + |f(0)| \leq rM'(r) + |f(0)|.$$

However,

$$f'(z) = \frac{1}{2\pi i} \int_C \frac{f(w)}{(w - z)^2}\, dw$$

where C is $|w - z| = R - r$ ($|z| = r < R$) and choosing z such that $|f'(z)| = M'(r)$

$$M'(r) \leq M(R)/(R - r).$$

We choose $R = 2r$ and

$$\frac{M(r) - |f(0)|}{r} \leq M'(r) \leq \frac{M(2r)}{r}.$$

Since

$$\varrho = \varlimsup_{r\to\infty} \frac{\log\log M(r)}{\log r}$$

the result follows. Note $\log M(r)$ is a convex function of $\log r$ and either $\log M(r)/\log r <$ constant A or $\to\infty$; therefore

$$\frac{\log\{\log M(2r) - \log r\}}{\log r} = \frac{\log\log M(2r)}{\log r} + \frac{\log\{1 - [\log r/\log M(2r)]\}}{\log r}.$$

If $\log M(r)/\log r \to \infty$, the second term tends to zero, and if $\log M(r)/\log r \to A > 1$, the second terms tends to zero also. Similarly for the left-hand side of the inequality. □

The "type" is discussed similarly.

The next theorem is of considerable consequence. Extensions and ramifications may be found in papers due to M. Marden.

5.7 Theorem (Laguerre). If $f(z)$ is an entire function, real for real z, of order $\varrho < 2$ and with real zeros, then the zeros of $f'(z)$ are also real. Further, the zeros of $f'(z)$ are separated from each other by zeros of $f(z)$.

Proof. The hypotheses give

$$f(z) = cz^k e^{az} \prod_{n=1}^{\infty} \left(1 - \frac{z}{z_n}\right) e^{z/z_n}$$

where k is zero or positive and c, a, $z_1, \ldots, z_n$ are all real. Taking logs and differentiating we obtain

$$\frac{f'(z)}{f(z)} = \frac{k}{z} + a + \sum_{n=1}^{\infty} \left(\frac{1}{z - z_n} + \frac{1}{z_n}\right).$$

If $z = x + iy$, then

$$\begin{aligned} \mathscr{I}\left\{\frac{f'(z)}{f(z)}\right\} &= -\, iy \left\{\frac{k}{x^2 + y^2} + \sum_{n=1}^{\infty} \frac{1}{(x - z_n)^2 + y^2}\right\} \\ &= 0 \quad \text{for} \quad y = 0 \quad \text{only.} \end{aligned}$$

Thus $f'(z) = 0$ on the real axis only. Also

$$\frac{d}{dz}\left\{\frac{f'(z)}{f(z)}\right\} = -\frac{k}{z^2} - \sum_{n=1}^{\infty} \frac{1}{(z - z_n)^2}$$

is real and negative for real z. Thus $f'(z)/f(z)$ decreases steadily as z increases through real values from z_n to z_{n+1} and it cannot vanish more than once between z_n and z_{n+1} (there can be no inflections). Clearly it changes sign (since $f = 0$ at z_n, z_{n+1} and f does not change sign between z_n, z_{n+1} but f' does (Fig. 2)). Consequently it vanishes exactly once in the interval and the theorem is proved. □

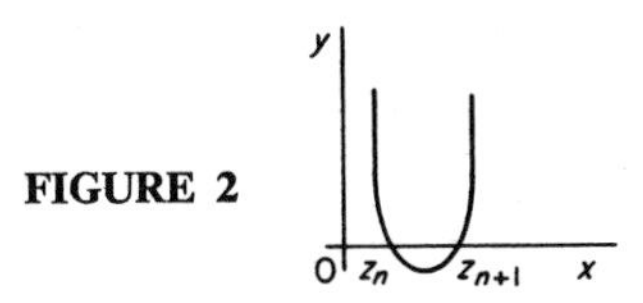

FIGURE 2

5.7.1 We can now deduce that if the zeros of $f'(z)$ are z_1', z_2', . . . , then the series $\sum_{n=1}^{\infty} 1/|z_n|^\alpha$, $\sum_{n=1}^{\infty} 1/|z_n'|^\alpha$ converge or diverge together. Thus the zeros of $f'(z)$ have the same exponent of convergence as those of $f(z)$.

It can be shown that $f(z)$ and $f'(z)$ have the same genus. The only case to consider is clearly when $\varrho = 1$ then the genus is 0 or 1. Since $f'(z)$ has the same order as $f(z)$ and has real zeros only, applying the same theorem to $f'(z)$, $f''(z)$ has real zeros only and so on for $f'''(z)$, etc.

We may extend the proof to functions of order 2 but of genus 1.† It is simple to see that the theorem is not true for functions of genus 2. Let

$$f(z) = z \exp z^2, \qquad f'(z) = (2z^2 + 1) \exp z^2$$

then the zeros of f are real and the zeros of f' are complex. Alternatively, let

$$f(z) = (z^2 - 4) \exp(z^2/3), \qquad f'(z) = \tfrac{2}{3}z(z^2 - 1) \exp(z^2/3)$$

and the zeros of f' are real but not separated by the zeros of f.

It is convenient at this stage to include Lucas's theorem on the zeros of a polynomial and its derivative. Extensions which have been deduced in various papers, require the techniques which have now been studied.

5.8 CONVEX SETS

We introduce the notion of convexity, and establish a theorem relating to the distribution of zeros of a polynomial and its derivative.

† See E. C. Titchmarsh, "The Theory of Functions." Oxford Univ. Press, London and New York, 1939.

5.8.1 Definition. A set R is convex whenever z_1 and z_2 are two points of R, and the points $[z \mid z = z_1 + \alpha(z_2 - z_1)]$, $0 \leq \alpha \leq 1$, also belong to R. Clearly a convex set must be arcwise connected, but need not be a domain (an open arcwise connected set.) The interior of a circle is a convex domain.

Also, the subset of S: $[z \mid z^2 - 1 \mid \leq 1, z \neq 0]$ (lemniscate) lying to the right or left of the half-plane is a convex region. Consider the interior of a triangle Δ with vertices at z_1, z_2, z_3. The interior is the point set,

$$\Delta_i \equiv \left[z \mid z = a_1 z_1 + a_2 z_2 + a_3 z_3;\ a_1 \geq 0,\ a_2 \geq 0,\ a_3 \geq 0;\ \sum_{r=1}^{3} a_r = 1 \right].$$

Similarly, Π is a convex polygon if Π is the boundary of a convex domain. If a polygon Π has vertices $z_1, z_2, \ldots, z_n$, the interior Π_i is such that

$$\Pi_i \equiv \left[z \mid z = a_1 z_1 + \cdots + a_n z_n;\quad a_1 \geq 0;\ \ldots;\ \sum_{r=1}^{n} a_r = 1 \right].$$

5.8.2 To prove: If $z = \sum_{r=1}^{n} a_r z_r$ and $a_r \geq 0$, $\sum_{r=1}^{n} a_r = 1$, then z lies in the convex hull of the z_r's.

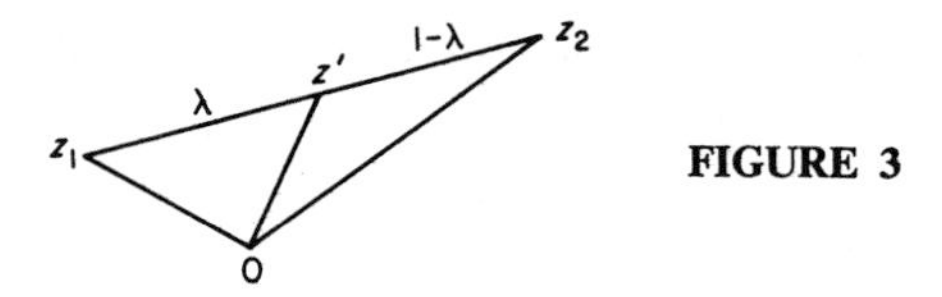

FIGURE 3

Proof (by induction). For the case $n = 2$, if z' is between z_1 and z_2 dividing the line joining z_1 and z_2 in the ratio $\lambda: 1 - \lambda$, $0 < \lambda < 1$ (Fig. 3), then vectorially

$$\overline{oz_1} = \overline{oz'} + \overline{z'z_1}$$

$$\overline{oz_2} = \overline{oz'} + \overline{z'z_2}$$

$$\overline{oz'} = \tfrac{1}{2}\{\overline{oz_1} + \overline{oz_2} + \overline{z_2z'} + \overline{z_1z'}\}$$

or

$$z' = \tfrac{1}{2}\{z_1 + z_2 + (1 - \lambda)(z_1 - z_2) + \lambda(z_2 - z_1)\}$$

and

$$z' = (1 - \lambda)z_1 + \lambda z_2\text{: } \sum \text{ coefficients} = 1.$$

Conversely, if $z' = a_1 z_1 + a_2 z_2$; $a_1 + a_2 = 1$; $a_1, a_2 \geq 0$, then z is in the convex hull of z_1, z_2. Let $z = a_1 z_1 + \cdots + a_n z_n$; $\sum_{r=1}^{n} a_r = 1$; $a_r \geq 0$.

We assume $0 < a_n < 1$, and

$$z = (1 - a_n)\left\{\frac{a_1}{1 - a_n} z_1 + \cdots + \frac{a_{n-1}}{1 - a_n} z_{n-1}\right\} + a_n z_n.$$

Write

$$z'' = \frac{a_1}{1 - a_n} z_1 + \cdots + \frac{a_{n-1}}{1 - a_n} z_{n-1}.$$

Since

$$\frac{a_1}{1 - a_n} + \cdots + \frac{a_{n-1}}{1 - a_n} = 1,$$

then z'' is in the convex hull of $z_1, \ldots, z_{n-1}$. Now $z = (1 - a_n)z'' + a_n z_n$ and since $a_n, 1 - a_n \geq 0$ and $\{a_n + (1 - a_n)\} = 1$, then z is in the convex hull of z'' and z_n, i.e., $z_1, \ldots, z_n$. □

The converse proposition is similarly demonstrated.

5.8.3 Lucas's Theorem. The zeros of the derivative $P'(z)$ of a polynomial $P(z)$, are contained within the convex hull of the zeros of $P(z)$.

Proof. Let $P(z)$ have zeros $z_1, z_2, \ldots, z_n$. Let Π be the least convex polygon containing these zeros. We show that $P'(z)$ cannot vanish anywhere in the exterior of Π.

Since $P(z) = (z - z_1)(z - z_2) \cdots (z - z_n)$, then

$$\frac{P'(z)}{P(z)} = \frac{d}{dz} \log P(z) = \sum_{k=1}^{n} \frac{1}{z - z_k}.$$

If $P'(z_0) = 0$, then

$$\sum_{k=1}^{n} 1/(z_0 - z_k) = 0 \quad \text{and} \quad \sum_{k=1}^{n} 1/(\bar{z}_0 - \bar{z}_k) = 0.$$

Thus

$$\sum_{k=1}^{n} (z_0 - z_k)/\mid z_0 - z_k \mid^2 = 0$$

and

$$z_0 \sum_{k=1}^{n} 1/\mid z_0 - z_k \mid^2 = \sum_{k=1}^{n} z_k/\mid z_0 - z_k \mid^2.$$

Since

$$z_0 = \sum_{r=1}^{n} a_r z_r \qquad \text{with} \quad \sum_{r=1}^{n} a_r = 1, \quad a_r \geq 0,$$

we deduce that z_0 lies within the convex hull of the z_r's. □

5.8.4 Some extensions of Lucas's theorem to entire functions can be found in a paper by M. Marden.[†] In this paper it is illustrated how Lucas's theorem is not true for arbitrary entire functions, e.g.,

$$f(z) = (z + 1)\exp(z^2/2)$$

has only one zero $z = -1$, but the derivative

$$f'(z) = (1 + z + z^2)\exp(z^2/2)$$

has two zeros; $z_1 = e^{2\pi i/3}$ and $z_2 = e^{4\pi i/3}$ and Lucas's theorem clearly does not hold. The theorem does extend to functions whose order ϱ is such that $0 \leq \varrho < 1$.

A few simple theorems which characterize meromorphic functions will now be included. The deeper structure of these functions is studied with the apparatus developed by Nevanlinna, Ahlfors, Hayman and others (see Hayman[‡]).

5.9 Theorem. Let $M_0(z)$ be a particular meromorphic function. If $G(z)$ is an arbitrary entire function, then

$$M(z) = M_0(z) + G(z)$$

is the most general meromorphic function which coincides with $M_0(z)$ in its poles and corresponding principal parts.

Proof. If $M_0(z)$ and $M(z)$ are two meromorphic functions which coincide in their poles and corresponding principal parts, their difference $M_0(z) - M(z)$ is evidently an entire function, say $G(z)$ and $M(z) = M_0(z) + G(z)$. □

For example, $\cot z$ and $2i/(e^{2iz} - 1)$ are two meromorphic functions which coincide in poles and corresponding principal part, thus they differ

[†] "On the zeros of the derivative of an entire function." *Amer. Math. Monthly* **25**, No. 8 (Oct. 1968).

[‡] W. K. Hayman, "Meromorphic Functions," Oxford Mathematical Monograph. Oxford Univ. Press (Clarendon) 1964. London and New York, 1964.

only by an additive entire function. The poles of $\cot z$ are the zeros of $\sin z = 0$, i.e., $z = n\pi$, $n = 0, \pm 1, \ldots$. The poles of $2i/(e^{2iz} - 1)$ are the roots of $e^{2iz} = 1 = e^{2n\pi i}$, i.e., $z = n\pi$, $n = 0, \pm 1, \ldots$. The difference is i, an entire function.

5.9.1 Lemma. Suppose $f(z)$ is analytic in the entire finite complex plane except for poles $z_1, z_2, \ldots, z_k$. Let $P(z, z_r)$ be the principal part of $f(z)$ at $z = z_r$, $r = 1, \ldots, k$. Then there exists a function $\phi(z)$, analytic in the entire finite complex plane such that

$$f(z) = \sum_{r=1}^{k} P(z, z_r) + \phi(z).$$

Moreover, $\phi(z)$ has the same principal part at $z = \infty$ as $f(z)$.

Proof. Consider

$$g(z) = \sum_{r=1}^{k} P(z, z_r), \qquad \text{where} \quad P(z, z_r) = \sum_{(j)} b_{jr}/(z - z_r)^j.$$

Clearly $g(z)$ is analytic in the entire finite complex plane except at points $z_1, \ldots, z_k$, where it has poles. Since for $s \neq r$, $P(z, z_s)$ is analytic at $z = z_r$, the principal part of $g(z)$ at $z = z_r$ is $P(z, z_r)$. Hence $g(z)$ has the same poles and principal parts at $z_1, \ldots, z_k$ as $f(z)$ and is analytic everwhere else. Consider $\phi(z) = f(z) - g(z)$, $z \neq z_r$, $r = 1, 2, \ldots$. This function is analytic in the entire complex plane except for possible poles at $z_1, z_2, \ldots, z_k$. Since $g(z)$ and $f(z)$ have the same poles and principal parts, the difference $\phi(z)$ is such that the principal part at each pole is zero. Thus for

$$f(z) = \frac{b_{-n}}{(z-a)^n} + \cdots + \frac{b_{-1}}{(z-a)} + b_0 + b_1(z-a) + \cdots,$$

$$g(z) = \frac{b_{-n}}{(z-a)^n} + \cdots + \frac{b_{-1}}{(z-a)},$$

$$\phi(a) = b_0,$$

and $\phi(z)$ has removable singularities at $z_1, \ldots, z_k$. Thus $\phi(z)$ is analytic. Since $g(z) \to 0$, $z \to \infty$, $g(z)$ has a removable singularity at $z = \infty$. Thus $\phi(z)$ has the same principal part at $z = \infty$ as $f(z)$. Writing

$$f(z) = g(z) + \phi(z),$$

the lemma is established. □

5.10 CHARACTERIZATION OF A MEROMORPHIC FUNCTION

A rational meromorphic function is determined to within an additive constant by the behavior at its poles. We want to find how completely an arbitrary meromorphic function is determined by the character of its poles. Weierstrass's theorem gives an infinite product representation of an entire function. We now prove the Mittag–Leffler theorem which represents a meromorphic function by an infinite decomposition into partial fractions.

Since a meromorphic function may be characterized by the nature of its poles only to within an added entire function and since the question has been answered for meromorphic functions with a finite number of poles, the only other possibility to answer is when ∞ is an accumulation point of poles. Clearly no finite point is an accumulation point of poles $(1/f \equiv 0)$.

5.10.1 Mittag–Leffler Theorem. Let $z_0, z_1, \ldots, z_n, \ldots$ be any sequence of distinct points tending to ∞. Suppose to each z_n there is associated a polynomial $P_n(1/(z-z_n))$ in the variable $1/(z-z_n)$. It is possible to find a meromorphic function $f(z)$ having poles at the points z_n but no other points, and with corresponding principal parts $P_n(1/(z-z_n))$. Then $f(z)$ may be represented in the form

$$f(z) = \omega(z) + \sum_{\nu=0}^{\infty} \{P_\nu(1/(z-z_\nu)) - q_\nu(z)\}$$

where $q_\nu(z)$ are polynomials and $\omega(z)$ is an entire function of z.

Proof. Unlike the finite case we must ensure that the given representation converges. Suppose the sequence $\{z_n\}$ is ordered. Thus $|z_0| \leq |z_1| \leq \cdots$ since the only point of accumulation is infinity. Possibly $z_0 = 0$, but all other points differ from zero. We suppose that $z_0 \neq 0$ and the function $P_\nu(1/(z-z_\nu))$ being analytic everywhere except at z_ν, must itself be analytic at the origin. Thus P_ν has a Taylor expansion at the origin given by

$$P_\nu(1/(z-z_\nu)) = a_0^{(\nu)} + a_1^{(\nu)}z + a_2^{(\nu)}z + \cdots$$

The radius of convergence is clearly $|z_\nu|$. The series converges uniformly in the circle C_ν: $|z| \leq \frac{1}{2}|z_\nu|$. Thus in C_ν, $P_\nu(1/(z-z_\nu))$ can be approximated by a finite sum as closely as we please. In particular, for $q_\nu(z) = a_0^{(\nu)} + a_1^{(\nu)}z + \cdots + a_{k_\nu}^{(\nu)}z^{k_\nu}$

$$|P_\nu(1/(z-z_\nu)) - q_\nu(z)| < 1/2^\nu$$

throughout C_ν. The series $\sum_{\nu=0}^{\infty}\{P_\nu(1/(z-z_\nu)) - q_\nu(z)\}$ converges to the desired meromorphic function in every circle about the origin since any such circle can be contained within one of the C_ν's.

In C_ν, $\sum_{n=0}^{\nu-1}\{P_n(1/(z-z_n)) - q_n(z)\}$ is well behaved. It is an analytic function with no singularities but the prescribed poles. Thus

$$\sum_{n=\nu}^{\infty}\{P_n(1/(z-z_n)) - q_n(z)\}$$

is analytic in C_ν and dominated by $\sum_{n=\nu}^{\infty}(1/2^n)$. The series $(\sum_{n=\nu}^{\infty})$ then converges uniformly in C_ν. Thus since a uniformly convergent series of analytic functions converges to an analytic function, the second part of the series introduces no new singularities into C_ν. Thus the theorem is proved (if $z_0 = 0$, we add on $P_0(1/z)$). □

5.10.2 In general, the polynomials $q_\nu(z)$ will not be uniformly bounded (to ensure convergence). However, in special circumstances all the $q_\nu(z)$ may be chosen of the same finite degree. It is sufficient to take the degree k_ν of the polynomial $q_\nu(z)$ [i.e., the sum of the first k_ν terms of the power series for $P_\nu(1/(z-z_\nu))$] so large, that having chosen an arbitrary $R > 0$, the terms $|P_\nu(1/(z-z_\nu)) - q_\nu(z)|$ of the series for all $|z| \leq R$ and large ν, remain less than the terms of a convergent series of positive terms.

EXAMPLE. The convergence producing terms $q_\nu(z)$ are not always necessary. If, e.g., the points $0, 1, 4, \ldots, \nu^2, \ldots$ are to be poles of order one with respective principal parts $1/(z-\nu^2)$, then

$$f(z) = \frac{1}{z} + \sum_{\nu=1}^{\infty}\frac{1}{z-\nu^2} = \sum_{\nu=0}^{\infty}\frac{1}{z-\nu^2}$$

is a solution. Note that for $R > 0$ and $m > \sqrt{2R}$, the series from $\nu = m+1$ with $|z| \leq R$, converges uniformly because

$$|1/(z-\nu^2)| \leq 1/(\nu^2 - R) < 1/(\nu^2 - \tfrac{1}{2}\nu^2) = 2/\nu^2.$$

Similarly if the function has only simple poles

$$P_\nu(1/(z-z_\nu)) = a_\nu/(z-z_\nu),$$

then the $q_\nu(z)$ may all be chosen of degree n if the series $\sum_{\nu=0}^{\infty}|a_\nu|/|z_\nu|^{n+2}$ converges. Since

$$\left|\frac{a_\nu}{z-z_\nu} - q_\nu(z)\right| = \left|-\frac{a_\nu}{z_\nu}\left(1 + \frac{z}{z_\nu} + \frac{z^2}{z_\nu^{\,2}} + \cdots\right) - q_\nu(z)\right|,$$

choosing $q_\nu(z)$ a polynomial of degree n, viz,

$$-(a_\nu/z_\nu)(1 + (z/z_\nu) + \cdots + (z^n/z_\nu{}^n)),$$

$$\left|\frac{a_\nu}{z - z_\nu} - q_\nu(z)\right| \leq \frac{|a_\nu|\,|z|^{n+1}}{|z_\nu|^{n+2}} + \frac{|a_\nu|\,|z|^{n+2}}{|z_\nu|^{n+3}} + \cdots.$$

Thus if $\sum_{\nu=0}^{\infty} |a_\nu|/|z_\nu|^{n+2}$ converges, it assures the convergence of $\sum_{(\nu)}((a_\nu/(z - z_\nu) - q_\nu(z))$. This case is the one that arises in most applications. Although the Mittag-Leffler theorem can be used to expand a function with simple poles into partial fractions, the entire function $\omega(z)$ has still to be determined. This special case has a more direct approach. Let C be any simple closed curve not passing through any pole of $f(z)$. If z is a regular point inside C, then

$$f(z) + \sum \operatorname{res}\{g(z)\} = \frac{1}{2\pi i}\int_C \frac{f(\xi)}{\xi - z}\,d\xi \quad \text{and} \quad g(z) = \frac{f(\xi)}{\xi - z},$$

i.e.,

$$f(z) = \frac{1}{2\pi i}\int_C \frac{f(\xi)}{\xi - z}\,d\xi - \sum_{(\nu)} \operatorname{res}\left\{\frac{f(z_\nu)}{z_\nu - z}\right\},$$

where $\operatorname{res}\{f(z_\nu)/(z_\nu - z)\}$ means residue of $f(\xi)/(\xi - z)$ at $\xi = z_\nu$, the sum taken over all singularities z_ν of $f(z)$ in C.

NOTE. $f(z)$ is the residue of $f(\xi)/(\xi - z)$ at the pole $\xi = z$.

Since $f(z)$ is assumed to have simple poles,

$$\operatorname{res}\left\{\frac{f(z_\nu)}{z_\nu - z}\right\} = \frac{\operatorname{res}\{f(z_\nu)\}}{z_\nu - z}$$

and

$$f(z) = \frac{1}{2\pi i}\int_C \frac{f(\xi)}{\xi - z}\,d\xi + \sum_{(\nu)} \frac{\operatorname{res}\{f(z_\nu)\}}{z - z_\nu}.$$

Since the poles are isolated $\exists$ a sequence of closed curves C_n such that $C_1 \supset C_2 \supset C_3 \supset \cdots \supset C_n \cdots$, each avoiding all the poles of $f(z)$ and such that the distance of C_n from the origin tends to infinity with n. If for some such sequence,

$$\lim_{n\to\infty} \int_{C_n} \frac{f(\xi)}{\xi - z}\,d\xi = 0,$$

denoting poles in the annulus between C_{n-1}, C_n by $z_\nu^{(n)}$, the series

$$f(z) = \sum_{n=1}^{\infty} \left\{ \sum_{(\nu)} \frac{\operatorname{res}\{f(z_\nu^{(n)})\}}{z - z_\nu^{(n)}} \right\}$$

converges and gives the decomposition of the function into partial fractions. As an example of the theorem and the above method we consider $\pi \cot \pi z$.

5.10.3 $\pi \cot \pi z$. The poles are the zeros of $\sin \pi z$ viz., poles of order 1 at $z = 0, \pm 1, \pm 2, \ldots$. The residue at $z = n$ is

$$\lim_{z \to n} (z - n) \cdot \pi \frac{\cos \pi z}{\sin \pi z} = 1.$$

Thus the principal parts

$$P_\nu\left(\frac{1}{z - z_\nu}\right) = \frac{1}{z - z_\nu} = -\frac{1}{z_\nu} - \frac{z}{z_\nu^2} - \cdots$$

$z_0 = 0$, $z_{2\nu-1} = \nu$, $z_{2\nu} = -\nu$, for $\nu = 1, 2, 3, \ldots$.

We may take the degree k_ν of the polynomials $q_\nu(z)$ to be zero, hence $q_\nu(z) = -1/z_\nu$ and

$$\left| P_\nu\left(\frac{1}{z - z_\nu}\right) - q_\nu(z) \right| = \left| \frac{1}{z - z_\nu} + \frac{1}{z_\nu} \right| = \frac{|z|}{|z_\nu|\,|z - z_\nu|} \leq \frac{R}{|z_\nu|\,|z - z_\nu|}.$$

Also, $|z - z_\nu| = |z_\nu - z| \geq |z_\nu| - |z| \geq |z_\nu| - R$. Thus

$$\frac{R}{|z_\nu|\,|z - z_\nu|} \leq \frac{R}{|z_\nu|(|z_\nu| - R)}$$

and for large $|z_\nu|$, viz., $|z_\nu| = \nu > 4R$,

$$R < \tfrac{1}{4}|z_\nu| \qquad \text{and} \qquad |z_\nu| - R > \tfrac{3}{4}|z_\nu|.$$

Therefore

$$\frac{R}{|z_\nu|\,|z_\nu - R|} < \frac{R}{|z_\nu|} \cdot \frac{4}{3|z_\nu|} < \frac{2R}{|z_\nu|^2}$$

and hence $|P_\nu - q_\nu|$ is less then the terms of a convergent series of positive terms. Thus

$$f(z) = \pi \cot \pi z = \omega(z) + \frac{1}{z} + \sum_{\nu=1}^{\infty} \left\{ \frac{1}{z - z_\nu} + \frac{1}{z_\nu} \right\}$$

or rewriting,

$$f(z) = \omega(z) + \frac{1}{z} + \sum_{n \neq 0} \left\{ \frac{1}{z - n} + \frac{1}{n} \right\}.$$

Further, $\omega(z)$ is an entire function and the sum is uniformly and absolutely convergent in any bounded region at positive distance from the integers or more precisely, on every set of the form

$$\{z \,\big|\, |z| \leq R,\ |z - n| \geq \delta > 0,\ n \in Z\}.$$

For this reason the terms of the series can be reordered to give $\pi \cot \pi z = \omega(z) + S(z)$, where

$$S(z) = \frac{1}{z} + \sum_{n=1}^{\infty} \frac{2z}{z^2 - n^2}.$$

(Note that $\sum_{n=1}^{\infty} 1/(z - n)$ is not convergent.) The problem of determining $\omega(z)$ can sometimes be difficult. If we know that

$$\sin \pi z = \pi z \prod_{n=1}^{\infty} \left(1 - \frac{z^2}{n^2}\right),$$

taking logs and differentiating, we have

$$\pi \cot \pi z = \frac{1}{z} + \sum_{n=1}^{\infty} \frac{2z}{z^2 - n^2}$$

showing that $\omega(z) = 0$. However, this is a rather circular argument and we consider the following. $S(z)$ is periodic with period 1, since

$$\begin{aligned} S(z) - S(z+1) &= \frac{1}{z} - \frac{1}{z+1} \\ &\quad + \lim_{M \to \infty} \sum_{n=1}^{M} \left\{ \frac{1}{z - n} + \frac{1}{z + n} - \frac{1}{z - (n-1)} - \frac{1}{z + (n+1)} \right\} \\ &= \lim_{M \to \infty} \left\{ \frac{1}{z - M} - \frac{1}{z + M + 1} \right\} = 0. \end{aligned}$$

Also, $S(z)$ is uniformly convergent in the region

$$\{z = x + iy \,\big|\, |y| \leq R,\ |z - n| \geq \delta > 0,\ n \in Z\}.$$

However, it is not convergent for unbounded y since if any ε is given satisfying $0 < \varepsilon < \frac{1}{2}$ and if an $N = N(\varepsilon)$ could then be found such that

$$\left| \sum_{n=N+1}^{\infty} \frac{2z}{z^2 - n^2} \right| < \varepsilon \qquad \text{for all } z,$$

putting $z = 2iN$ we would have

$$\left| \sum_{n=N+1}^{\infty} \frac{2z}{z^2 - n^2} \right| = \left| (-i) \sum_{n=N+1}^{\infty} \frac{4N}{n^2 + 4N^2} \right|$$

$$> \sum_{n=N+1}^{2N} \frac{4N}{n^2 + 4N^2} \geq \sum_{n=N+1}^{2N} \frac{4N}{4N^2 + 4N^2} = \frac{1}{2}.$$

Then $\pi \cot \pi z$ is periodic with period 1 and so $\omega(z)$ must be also periodic with period 1. Further

$$\pi \cot \pi z = \pi \frac{\cos \pi x \cosh \pi y - i \sin \pi x \sinh \pi y}{\sin \pi x \cosh \pi y + i \cos \pi x \sinh \pi y}$$

$$\to \pm \pi \frac{e^{-i\pi x}}{ie^{-i\pi x}} = \mp i\pi \qquad \text{as} \quad y \to \pm \infty$$

and thus $\pi \cot \pi z$ is bounded as $|y| \to +\infty$. Similarly $|S(z)|$ is bounded as $|y| \to +\infty$, since using $|z^2 - n^2|^2 - (n^2 + y^2 - 1)^2 \geq 0$ or observing that $|f| \geq -\mathscr{R}\{f\}$,

$$\left| \frac{2z}{z^2 - n^2} \right| \leq \frac{2(|y| + 1)}{n^2 + y^2 - 1} \qquad \text{for} \quad z \in \{z \mid 0 \leq x \leq 1\}$$

and

$$\sum_{n=1}^{\infty}{}' \frac{2(y+1)}{n^2 + y^2 - 1}$$

$$\leq \sum_{n<y} \frac{2(y+1)}{y^2 - 1} + \sum_{n \geq y} \frac{2(y+1)}{n^2 - n} \leq 2 + \frac{2(y+1)}{y-1} \leq 8$$

$$\text{for} \quad y \geq 2.$$

Note also that $S(iy) = -i\pi \coth \pi y \to \mp i\pi$ as $y \to \pm\infty$. Thus $\omega(z)$ is periodic with period 1 and bounded as $|y| \to \infty$, hence it is bounded in $\{z = x + iy \mid 0 \leq x \leq 1\}$ and so is bounded in the whole plane. However, it is an entire function, thus it is a constant. Since $\omega(-z) = -\omega(z)$, we have that $\omega(z) = 0$ for all z [or examine $\omega(iy)$]. Hence

$$\pi \cot \pi z = \frac{1}{z} + \sum_{n=1}^{\infty} \frac{2z}{z^2 - n^2}.$$

Alternatively, let

$$I_n = \frac{1}{2\pi i} \int_{R_n} \frac{\pi \cot \pi z}{z - \xi} \, dz.$$

Let R_n be the square $|\mathscr{R}(z)| = n + \frac{1}{2}$, $|\mathscr{I}(z)| = n + \frac{1}{2}$, $n = 0, \pm 1, \ldots,$ the vertical sides passing between the poles at n, $n + 1$

$$I_n = \frac{1}{4\pi i}\left\{\int_{R_n} \frac{\pi \cot \pi z}{z - \xi}\, dz + \int_{R_n} \frac{\pi \cot \pi(-z)}{-z-\xi}\, d(-z)\right\}$$

$$= \frac{1}{4\pi i}\int_{R_n} \pi \cot \pi z\left(\frac{1}{z-\xi} - \frac{1}{z+\xi}\right) dz$$

$$= \frac{1}{4\pi i}\int_{R_n} \frac{2\xi}{z^2 - \xi^2}\, \pi \cot \pi z\, dz.$$

$$\left|\frac{1}{2\pi i}\int_{R_n} \frac{\pi \cot \pi z}{z-\xi}\, dz\right|$$

$$\leq \frac{1}{4\pi}\cdot (\text{bound for } |\pi \cot \pi z| \text{ on } R_n)\cdot \frac{2|\xi|}{(n+\frac{1}{2})^2 - |\xi|^2}\, 8\left(n + \frac{1}{2}\right)$$

$$\to 0 \quad \text{as} \quad n \to \infty$$

and $\pi \cot \pi\xi = S(\xi)$.

CHAPTER VI

FUNCTIONS WITH REAL AND/OR NEGATIVE ZEROS: MINIMUM MODULUS I AND SEQUENCES OF FUNCTIONS

6.1 FUNCTIONS WITH REAL ZEROS ONLY

Some important complex functions have real zeros only, e.g., $1/\Gamma(z)$ has no complex zeros. However, it can be very difficult sometimes to decide whether the zeros are real or not.† The question can be decided sometimes.

6.1.1 Theorem. Let $f(z)$ be a polynomial with

$$f(z) = a_0 + \cdots + a_p z^p$$

all of whose zeros are real. Let $\phi(\omega)$ be an entire function of genus 0 or 1, all zeros being real and negative. Then

$$g(z) = a_0\phi(0) + a_1\phi(1)z + \cdots + a_p\phi(p)z^p$$

has all its zeros real, with as many positive, zero, and negative zeros as $f(z)$.

† Cf. E. C. Titchmarsh, "The Theory of Functions," Section 8.6, p. 268. Oxford Univ. Press, London and New York, 1939.

Proof. Let

$$\phi(\omega) = ae^{k\omega} \prod_{n=1}^{\infty} \left(1 + \frac{\omega}{\alpha_n}\right) \exp(-\omega/\alpha_n), \qquad \alpha_n > 0 \quad \text{for} \quad \text{all } n.$$

Consider

$$\begin{aligned} g_1(z) &= f(z) + \frac{z}{\alpha_1} f'(z) \\ &= \left(1 + \frac{\theta}{\alpha_1}\right) f(z) \qquad \left(\text{where} \quad \theta \equiv z \frac{\partial}{\partial z}\right) \\ &= \frac{z^{1-\alpha_1}}{\alpha_1} \frac{d}{dz} (z^{\alpha_1} f(z)) \qquad (z > 0) \\ &= a_0 + a_1\left(1 + \frac{1}{\alpha_1}\right) z + \cdots + a_p\left(1 + \frac{p}{\alpha_1}\right) z^p. \end{aligned}$$

Then $g_1(z)$ has as many zeros at $z = 0$ as $f(z)$ since $g_1(0) = f(0)$. Putting $h(z) = z^{\alpha_1} f(z)$, $\alpha_1 > 0$, the zeros of h are $z = 0$ and the zeros of f. Hence, if $h(z)$ vanishes, e.g., at $z = b_1, b_2$ [zeros of $f(z)$], then

$$h'(z_1) = 0 \qquad \text{for} \quad b_1 < z_1 < b_2 \quad \text{(Rolle's theorem).}$$

Thus $g_1(z)$ has the same number of positive and negative zeros as does f. [Note $g_1(z)$ is a polynomial of degree p and so is $f(z)$.] Since $\alpha_1 > 0$ and f is a polynomial, $z^{\alpha_1} f \to 0$ as $z \to 0$, hence there is another zero z_1' between 0 and z_1. Thus the operator

$$\left(1 + \frac{\theta}{\alpha_1}\right) \equiv \left(1 + \frac{z}{\alpha_1} \frac{\partial}{\partial z}\right)$$

cannot decrease the number of real positive zeros (it may increase them). The same result holds for

$$g_n(z) = a_0 + a_1\phi_n(1)z + \cdots + a_p\phi_n(p)z^p$$

where

$$\phi_n(\omega) = \left(1 + \frac{\omega}{\alpha_1}\right)\left(1 + \frac{\omega}{\alpha_2}\right) \cdots \left(1 + \frac{\omega}{\alpha_n}\right),$$

for then

$$g_n(z) = \prod_{r=1}^{n} \left(1 + \frac{\theta}{\alpha_r}\right) f(z).$$

Transforming $z = z' \exp k_n$, $k_n = k - \sum_{\nu=1}^{n} 1/\alpha_\nu$, we have

$$g_n(z' \exp k_n) = a_0 + a_1\phi_n(1)z' \exp\left(k - \sum_\nu \frac{1}{\alpha_\nu}\right) + \cdots + a_p\phi_n(p)z'^p \exp\left(kp - p\sum_\nu \frac{1}{\alpha_\nu}\right)$$

and

$$ag_n(z' \exp k_n) = aa_0 + aa_1\phi_n(1)z' \exp\left(k - \sum_\nu \frac{1}{\alpha_\nu}\right) + \cdots \equiv G_n(z).$$

Thus

$$G_n(z) = a_0\Phi_n(0) + \cdots + a_p\Phi_n(p)z^p$$

where

$$\begin{aligned}\Phi_n(\omega) &= a \exp(k_n\omega)\,\phi_n(\omega)\\ &= a\left\{\exp\left(k\omega - \frac{\omega}{\alpha_1} - \cdots - \frac{\omega}{\alpha_n}\right)\right\}\left(1 + \frac{\omega}{\alpha_1}\right)\cdots\left(1 + \frac{\omega}{\alpha_n}\right)\\ &= a \exp(k\omega) \prod_{\nu=1}^{n}\left(1 + \frac{\omega}{\alpha_\nu}\right)\exp\left(-\frac{\omega}{\alpha_\nu}\right).\end{aligned}$$

Thus $\Phi_n(\omega) \to \phi(\omega)$ uniformly in any finite region, since ϕ is an entire function, and by Hurwitz's theorem, the zeros of $g(z)$ are limits of zeros of $G_n(z)$. Then $g_n(z)$ has at least as many positive and negative zeros as $f(z)$, also, $g(z)$ has as many zeros at $z = 0$ as $f(z)$ since if $g(z)$ has a zero of order r ($\leq p$) then $a_0, a_1, \ldots, a_{r-1} = 0$. Thus $f(z)$ has a zero of order r. (Note this has nothing to do with Rolle's theorem.) Hence the theorem is proved. □

A direct consequence is as follows.

6.1.2 Theorem. Suppose that $\phi(\omega)$ satisfies the conditions of the previous theorem and that $f(z)$ is an entire function of the form

$$f(z) = e^{az+b}\prod_{n=1}^{\infty}\left(1 + \frac{z}{z_n}\right), \qquad a \text{ and } z_n \text{ all positive.}$$

Let $f(z) = \sum_{n=0}^{\infty} a_n z^n$. Then

$$g(z) = \sum_{n=0}^{\infty} a_n\phi(n)z^n$$

is an entire function all of whose zeros are real and negative.

Proof. Since $(1+x)e^{-x} \leq 1$, $x \geq 0$, $g(z)$ is an entire function. Thus $|\phi(n)| \leq |a|\, e^{kn}$ and

$$\lim_{n\to\infty} \sqrt[n]{|a_n\phi(n)|} \leq \lim_{n\to\infty} \sqrt[n]{|a_n|} \cdot |a|^{1/n} e^k = 0$$

since $f(z)$ is an entire function. Hence the $g(z)$ series converges everywhere. Let

$$f_p(z) = e^b\left(1 + \frac{az}{p}\right)^p \prod_{n=1}^{p}\left(1 + \frac{z}{z_n}\right)$$
$$= \sum_{n=0}^{2p} a_{n,p} z^n,$$

a polynomial of degree $2p$. All zeros of $f_p(z)$ are real and negative. Thus by the previous theorem so are the zeros of

$$g_p(z) = \sum_{n=0}^{2p} a_{n,p}\phi(n) z^n.$$

We show that $g_p(z) \to g(z)$ uniformly in any finite region. Since

$$\lim_{p\to\infty} f_p(z) = f(z) \qquad \text{as} \qquad \left(1 + \frac{az}{p}\right)^p \to e^{az},$$

then $a_{n,p} \to a_n$ as $p \to \infty$ for fixed n. Also $|a_{n,p}| \leq |a_n|$ for all n, p since

$$\left(1 + \frac{az}{p}\right)^p = 1 + az + \left(1 - \frac{1}{p}\right)\frac{a^2z^2}{2!} + \cdots + \frac{a^p z^p}{p^p}.$$

Thus for $2p > N$,

$$|g(z) - g_p(z)|$$
$$= \left|\sum_{n=0}^{N} a_n z^n \phi(n) + \sum_{n=N+1}^{\infty} a_n z^n \phi(n) - \sum_{n=0}^{N} a_{n,p} z^n \phi(n) - \sum_{n=N+1}^{2p} a_{n,p} z^n \phi(n)\right|$$
$$\leq \left|\sum_{n=0}^{N} (a_n - a_{n,p}) z^n \phi(n)\right| + \left|\sum_{n=N+1}^{\infty} a_n z^n \phi(n)\right| + \left|\sum_{n=N+1}^{2p} a_{n,p} z^n \phi(n)\right|$$
$$\leq \left|\sum_{n=0}^{N} (a_n - a_{n,p}) z^n \phi(n)\right| + 2 \sum_{n=N+1}^{\infty} |a_n z^n \phi(n)|.$$

We can choose N so large that the second term is less than ε [since $g(z)$ is entire]. Then for this fixed N the first term tends to zero and $g_p(z) \to g(z)$. As in the previous theorem, the result follows from Hurwitz's theorem. □

EXAMPLE 1. For $f(z) = e^z$, if $\phi(\omega)$ satisfies the conditions of the theorem, then $F(z) = \sum_{n=0}^{\infty}[\phi(n)/n!]\, z^n$ is an entire function with all zeros real and negative.

EXAMPLE 2. For

$$\phi(\omega) = \frac{1}{\Gamma(\omega + \nu + 1)}, \qquad \nu > -1,$$

this is an entire function of genus 1 with zeros at $\omega = -\nu - 1, -\nu - 2, \ldots$ all real and negative. Thus the zeros of

$$\sum_{n=0}^{\infty} (z^n/n!)\Gamma(n + \nu + 1) = J_\nu(2i\sqrt{z})/(i\sqrt{z})^\nu$$

are all real and negative and the zeros of $J_\nu(z)$ are all real. [Take $z = -a$. Then $2i\sqrt{z} = -2a^{1/2}$ which is real.]

6.1.3 Functions with Real Negative Zeros. If all the zeros of a function are real and negative, the modulus of the function is related to the distribution of the zeros in a very simple way. Let $f(z)$ be such a function with order $\varrho < 1$. Then

$$f(z) = \prod_{n=1}^{\infty}\left(1 + \frac{z}{z_n}\right).$$

If z is real and greater than 0, then

$$\begin{aligned}\log f(z) &= \sum_{n=1}^{\infty} \log\left(1 + \frac{z}{z_n}\right)\\ &= \sum_{n=1}^{\infty} n\left\{\log\left(1 + \frac{z}{z_n}\right) - \log\left(1 + \frac{z}{z_{n+1}}\right)\right\}\\ &= \sum_{n=1}^{\infty} n \int_{z_n}^{z_{n+1}} \frac{z\,dt}{t(z+t)}\\ &= z \int_0^{\infty} \frac{n(t)}{t(z+t)}\, dt.\end{aligned}$$

(The first integral step depends upon the z_n being real, greater than 0, and ordered increasingly.)

6.1.4 Theorem. If as $t \to \infty$, $n(t) \sim \lambda t^\varrho (\lambda > 0)$ (excludes $\varrho = 0$ except for f a polynomial), then $\log f(x) \sim \pi\lambda x^\varrho \csc \pi\varrho$, $x \to \infty$, where all the zeros of f are real and negative, $f(0) = 1$, f is entire, and of order $0 < \varrho < 1$.

Proof. We have $(\lambda - \varepsilon)t^{\varrho} < n(t) < (\lambda + \varepsilon)t^{\varrho}$ for $t > t_0(\varepsilon)$. Thus

$$\begin{aligned}\log f(x) &< x \int_0^{t_0} \frac{n(t)}{t(x+t)}\,dt + x \int_{t_0}^{\infty} \frac{(\lambda+\varepsilon)t^{\varrho}}{t(x+t)}\,dt \\ &= x \int_0^{t_0} \frac{n(t) - (\lambda+\varepsilon)t^{\varrho}}{t(x+t)}\,dt + x \int_0^{\infty} \frac{(\lambda+\varepsilon)t^{\varrho}}{t(x+t)}\,dt \\ &= O(1) + x \int_0^{\infty} \frac{(\lambda+\varepsilon)t^{\varrho}}{t(x+t)}\,dt\end{aligned}$$

since

$$\begin{aligned}\left| x \int_0^{t_0} \frac{n(t) - (\lambda+\varepsilon)t^{\varrho}}{t(x+t)}\,dt \right| &\leq |x| \int_0^{t_0} \frac{|n(t) - (\lambda+\varepsilon)t^{\varrho}|}{t\,|x|}\,dt \\ &= \int_0^{t_0} |n(t) - (\lambda+\varepsilon)t^{\varrho}| \frac{dt}{t}\end{aligned}$$

which is independent of x. Also, the integral converges at the lower limit since $n(t) = 0$ for $t < z_1$. Transforming $t = xu$ in the second integral, we obtain

$$x^{\varrho}(\lambda+\varepsilon) \int_0^{\infty} \frac{u^{\varrho-1}}{1+u}\,du = x^{\varrho}(\lambda+\varepsilon)\pi \csc \pi\varrho, \quad 0 < \varrho < 1.$$

A similar result holds for $\lambda - \varepsilon$ and the theorem follows. □

Generally,

$$\log f(re^{i\theta}) \sim e^{i\varrho\theta}\pi\lambda \csc \pi\varrho \cdot r^{\varrho}$$

for fixed $\theta \in (-\pi, \pi)$, $\log f(z)$ denoting the branch which is real on the positive real axis. In fact the expression for $\log f(z)$ as an integral for real z, holds by analytic continuation for $-\pi < \arg z < \pi$ (excludes real negative axis), and as before

$$\log f(re^{i\theta}) \sim re^{i\theta} \int_0^{\infty} \frac{\lambda t^{\varrho}}{t(re^{i\theta} + t)}\,dt.$$

Turning the line of integration to $t = ue^{i\theta}$,

$$re^{i\theta} \int_0^{\infty} \frac{\lambda t^{\varrho}}{t(re^{i\theta}+t)}\,dt = \lambda re^{i\varrho\theta} \int_0^{\infty} \frac{u^{\varrho}}{u(r+u)}\,du = \lambda r^{\varrho} e^{i\varrho\theta}\pi \csc \pi\varrho$$

as before.

6.1.5 Conversely it can be shown that if as $x \to \infty$ though real values $\log f(x) \sim \pi\lambda x^{\varrho} \csc \pi\varrho$, $(0 < \varrho < 1, \lambda > 0)$ then $n(r) \sim \lambda r^{\varrho}$. This theorem is more difficult than the previous one. A clear proof is given by Boas,[†] and a lengthy proof is given by Titchmarsh.[‡]

Several extensions and further results concerning functions with real and/or negative zeros may be found in Boas's book.[†]

6.2 THE MINIMUM MODULUS

Let $m(r)$ denote the minimum modulus of $|f(z)|$ on $|z| = r$. We cannot expect $m(r)$ to behave as simply as $M(r)$ since it vanishes whenever r is the modulus of a zero of $f(z)$. Except in the immediate neighborhood of these exceptional points, a lower limit can be set for $m(r)$. Generally $m(r) \to 0$ in somewhat the same way as $1/M(r)$. Note $m(r)$ can be quite large for certain functions. Consider a canonical product $P(z)$ of order ϱ, with zeros $z_1, z_2, \ldots, z_n, \ldots$.

6.2.1 Theorem. About each zero z_n $(|z_n| > 1)$, we describe a circle of radius $1/r_n^h$, $h > \varrho$. Then in the region excluded from these circles, $|P(z)| > \exp(-r^{\varrho+\varepsilon})$, $r > r_0(\varepsilon)$, $\varepsilon > 0$.

Proof. By Theorem 4.10

$$\log|P(z)| = \sum_{r_n \le kr} \log\left|1 - \frac{z}{z_n}\right| + \sum_{r_n \le kr} \left|\frac{z}{z_n} + \cdots + \frac{1}{p}\left(\frac{z}{z_n}\right)^p\right|$$

$$+ \sum_{r_n > kr} \log\left\{\left|1 - \frac{z}{z_n}\right| \left|\exp\left(\frac{z}{z_n} + \cdots + \frac{1}{p}\left(\frac{z}{z_n}\right)^p\right)\right|\right\}$$

$$\ge \sum_{r_n \le kr} \log\left|1 - \frac{z}{z_n}\right| - \sum_{r_n \le kr} O\left(\frac{r}{r_n}\right)^p - \sum_{r_n > kr} O\left(\frac{r}{r_n}\right)^{p+1}$$

since

$$\sum_{r_n \le kr} \left(\frac{r}{r_n}\right)^p = \sum_{r_n \le kr} \left(\frac{r}{r_n}\right)^{\varrho+\varepsilon} \left(\frac{r_n}{r}\right)^{\varrho+\varepsilon-p} \le k^{\varrho+\varepsilon-p} \sum_{r_n \le kr} \left(\frac{r}{r_n}\right)^{\varrho+\varepsilon}$$

$$\le k^{\varrho+\varepsilon-p} r^{\varrho+\varepsilon} \sum_{n=1}^{\infty} r_n^{-\varrho-\varepsilon}$$

$$= O(r^{\varrho+\varepsilon})$$

† R. Boas, "Entire Functions." Academic Press, New York, 1954.

‡ E. C. Titchmarsh, "On integral functions with real negative zeros," **26** (1927) 185–200. *Proc. London Math. Soc.* (2).

and

$$\sum_{r_n>kr}\left(\frac{r}{r_n}\right)^{p+1}=\sum_{r_n>kr}\left(\frac{r}{r_n}\right)^{\varrho+\varepsilon}\left(\frac{r}{r_n}\right)^{p+1-\varrho-\varepsilon}\leq k^{\varrho+\varepsilon-p-1}r^{\varrho+\varepsilon}\sum_{n=1}^{\infty}r_n^{-\varrho-\varepsilon}$$
$$=O(r^{\varrho+\varepsilon}).$$

We have assumed that $\varrho+\varepsilon-p>0$ and that $p+1-\varrho-\varepsilon\geq 0$. This fails if $\varrho=p+1$ but we then only need $\sum_{(n)}r_n^{-\varrho}=\sum_{(n)}r_n^{-p-1}<\infty$ and in this case we replace ε by 0 throughout. Thus

$$\log|P(z)|\geq\sum_{r_n\leq kr}\log\left|1-\frac{z}{z_n}\right|-O(r^{\varrho+\varepsilon}).$$

Further, since $\sum_{(n)}1/r_n^{h}<\infty$, the sum of the radii of the circles is finite (radius $R_n=1/r_n^{h}$). Thus $\exists$ circles centered at the origin and of arbitrarily large radius which lie entirely in the excluded region.

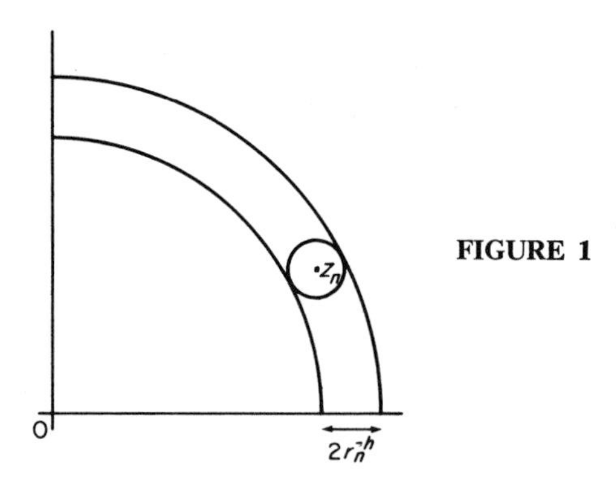

FIGURE 1

NOTE. The disk centered at z_n and with radius r_n^{-h} in Fig. 1 is confined to the annulus $r_n-r_n^{-h}<|z|<r_n+r_n^{-h}$. This meets the real axis (or any ray from the origin) in an interval of length $2r_n^{-h}$. These intervals are to be excluded. The intervals may overlap but the total length of the excluded intervals is less than or equal to $\sum_{n=1}^{\infty}2r_n^{-h}=S$ and $S<\infty$. If we now consider

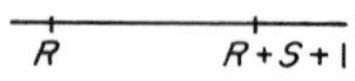

R —— $R+S+1$

FIGURE 2

any range of r of length greater than S, say, $R\leq r\leq R+S+1$ (Fig. 2), then the excluded intervals of r in this range have total length less than or equal to S so they do not cover the whole range. Thus $\exists r$ in this range not excluded and $r\geq R$ is arbitrarily large. If z lies outside every circle with center z_n and radius r_n^{-h}, since $r_n \leq kr$, then

$$\left|1 - \frac{z}{z_n}\right| > r_n^{-1-h} > (kr)^{-1-h}.$$

Hence

$$\sum_{1 < r_n \leq kr} \log\left|1 - \frac{z}{z_n}\right| > -(1+h)\log kr \cdot n(kr)$$

$$> - K \log kr \cdot r^{\varrho+\varepsilon} > - r^{\varrho+2\varepsilon}$$

(Note $r^{\varepsilon} > A \log r$.) Also,

$$\sum_{r_n \leq 1} \log\left|1 - \frac{z}{z_n}\right| > 0, \qquad r > 2.$$

Since for $r_n \leq 1$, $|z/z_n| > 2$,

$$\left|\frac{z}{z_n} - 1\right| \geq \left|\frac{z}{z_n}\right| - 1 > 1$$

and

$$\log\left|\frac{z}{z_n} - 1\right| > 0.$$

Thus

$$\log |P(z)| > - r^{\varrho+2\varepsilon} - O(r^{\varrho+\varepsilon})$$

and

$$|P(z)| > \exp(- r^{\varrho+\varepsilon}). \quad \square$$

6.2.2 The result now follows that if f is a function of order ϱ, then

$$m(r) \neq o(\exp(- r^{\varrho+\varepsilon})) \qquad \text{for any } \varepsilon > 0.$$

It was conjectured that for functions of finite order, $m(r) \neq o\{M(r)\}^{-1-\varepsilon}$, $\varepsilon > 0$, however, this was disproved by Hayman.† For $f(z) = P(z)e^{Q(z)}$, where $Q(z)$ is a polynomial of degree $q \leq \varrho$,

$$|\exp Q(z)| \geq \exp(- Ar^{q}) \geq \exp(- Ar^{\varrho}) \qquad \text{for large } r.$$

† W. K. Hayman, "The minimum modulus of large integral functions," *Proc. London Math. Soc.* (3) **2** (1952) 469–512.

Also, by the previous result [with ϱ_1 the order of the canonical product $P(z)$],

$$|P(z)| > \exp(-r^{\varrho_1+\varepsilon'}) \geq \exp(-r^{\varrho+\varepsilon'}).$$

Thus

$$|f(z)| \geq \exp(-r^{\varrho}[r^{\varepsilon'} + A])$$

and $m(r) \geq \exp(-r^{\varrho+\varepsilon})$, ε suitably adjusted.

6.3 THEOREMS ON SEQUENCES OF FUNCTIONS

We commence this section with the well-known Heine–Borel theorem. The theorems that follow all deal with the concept of compactness. We present the results without further amplification since they are somewhat sequential and quite self-explanatory.

6.3.1 Definition. A set or region is *compact* if it is both bounded and closed, e.g., (1) the set of all complex numbers z, such that $|z| \leq k$ (constant) is a compact set. However, (2) the set of all real numbers in $I = (0, 1)$ is not compact since it is bounded but not closed.

6.3.1.1 Theorem. (Heine–Borel). Let S be a compact set. Suppose there is a family $\{G_\alpha\}$ of open sets such that each point of S is contained in at least one of the G_α. Then there exists a finite subfamily $\{G_{\alpha_j}\}$ $(j = 1, 2, \ldots, n)$ of $\{G_\alpha\}$ such that every point of S is in at least one of the G_{α_j}.

Proof. Suppose that no finite subfamily of $\{G_\alpha\}$ covers S. (That is, there is no finite subfamily $\{G_{\alpha_j}\}$, $j = 1, 2, \ldots, n$ of the G_α such that every point of S is in at least one of the G_{α_j}.) Since S is a bounded set, it is contained in some closed square Q whose sides have length k. Subdivide Q into four closed congruent squares, the length of whose sides is $k/2$. Then there must be at least one of these, say Q_1, such that no finite subfamily of $\{G_\alpha\}$ covers $S \cap Q_1$ (that is the part of S contained in Q_1). Subdividing Q_1 into four closed congruent squares with sides of length $k/2^2$, for at least one of these squares, denoted by Q_2, there is no finite subfamily of $\{G_\alpha\}$ which covers $S \cap Q_2$.

Continuing this process we obtain an infinite nested sequence of closed squares $Q \supset Q_1 \supset \cdots$ such that diameter $(Q_n) \to 0$ and no finite subfamily of $\{G_\alpha\}$ covers $S \cap Q_n$. There is a point z_0 common to all the squares Q_n. This point is in S and hence contained in one of the sets of $\{G_\alpha\}$, say G_{α_p}. Since G_{α_p} is an open set, there exists an $\varepsilon > 0$ such that all z satisfying

$|z - z_0| < \varepsilon$ are contained in G_{α_p}. Furthermore, since z_0 is contained in all the Q_n and diameter $(Q_n) \to 0$, it follows that for n sufficiently large, $|z - z_0| < \varepsilon$ for every z in Q_n and thus Q_n is contained in G_{α_p}. Hence G_{α_p} covers $S \cap Q_n$. This contradicts the condition that no finite subfamily of $\{G_\alpha\}$ covers the part of S contained in any Q_n. Thus the theorem is established. □

6.3.2 Lemma. If (i) $\{f_n(z)\}$ is a sequence of functions each analytic in a domain D, (ii) $f_n(z) \to f(z)$ uniformly as $n \to \infty$, in every compact region $R \subset D$, (iii) $f(z)$ is not constant in D and (iv) $f(z) = a$ at some point $z_0 \in D$, then all but a finite number of the functions $f_n(z)$ take the value a in D.

NOTE. The result is not necessarily true if (iii) is omitted, e.g., $f_n(z) = z/n$ $(n = 1, 2, \ldots)$, $f(z) = 0$. Then $f_n(z) \to f(z)$ uniformly in every compact region as $n \to \infty$. However, if D does not contain $z = 0$, then $f_n(z) \neq 0$ for all $z \in D$.

Proof. Clearly $f(z)$ is analytic in D. Take $\varrho > 0$, such that $z \in D$ when $|z - z_0| \leq \varrho$ (Fig. 3) and $f(z) \neq a$ for $0 < |z - z_0| \leq \varrho$ [possible by (iii)]. Let R be a compact region $|z - z_0| \leq \varrho$ and C the circle

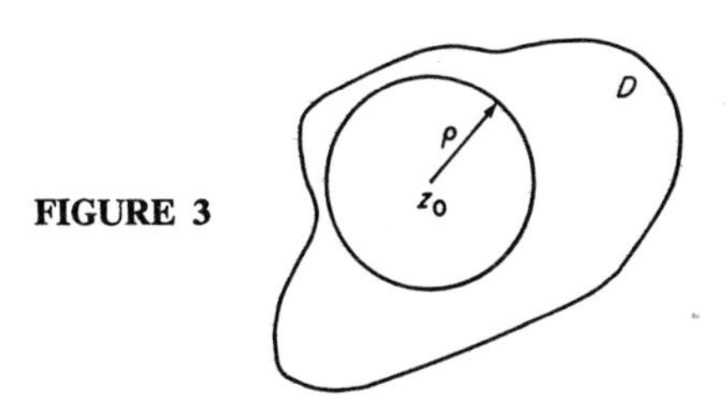

FIGURE 3

$|z - z_0| = \varrho$. Write $m = \underline{\lim}_{z \in C} |f(z) - a|$, then $m > 0$ since C is a compact set. Since $f_n(z) \to f(z)$ uniformly in R as $n \to \infty$, $\exists \nu$ such that $|f_n(z) - f(z)| < m$ for $n > \nu$ and all $z \in R$, hence

$$|f_n(z) - f(z)| < |f(z) - a| \qquad \text{for} \quad n > \nu \quad \text{and} \quad \text{all } z \in C.$$

Also if $f(z)$ is analytic in D, by (i) and (ii), so are $f_n(z) - f(z)$ and $f(z) - a$. Hence by Rouché's theorem, if $n > \nu$,

$$f_n(z) - f(z) + f(z) - a = f_n(z) - a$$

and $f(z) - a$ have the same number of zeros within C. Thus if $n > \nu$,

$f_n(z) = a$ at some point z within C and hence at some point $z \in D$ (*all but a finite number* comes from $n > \nu$). □

6.3.3 Definition. If a function $\omega = f(z)$ is single-valued in a region M and its inverse $z = \phi(\omega)$ is single-valued in N (the set of points ω corresponding to all possible points z of M), the transformation $\omega = f(z)$ is one–one or biuniform in M and $f(z)$ is *univalent* or *schlicht* in M, i.e., under a one–one transformation $\omega = f(z)$, any two distinct points of M map onto two distinct points of N, i.e., two points cannot "merge."

6.3.3.1 Theorem. If (i) $\{f_n(z)\}$ is a sequence of functions analytic in a domain D, (ii) $f_n(z) \to f(z)$, $n \to \infty$, uniformly in every compact region $R \subset D$, (iii) $f(z)$ is not constant in D, (iv) each $f_n(z)$ is univalent in D, then $f(z)$ is univalent in D.

Proof. Suppose $f(z)$ is not univalent in D. Then $\exists$ points z_1, z_2 in D such that $f(z_1) = f(z_2) = k$, for example. Let V_1, V_2 be neighborhoods of z_1, z_2, respectively (Fig. 4), such that $V_1 \subset D$, $V_2 \subset D$, and $V_1 \cap V_2 = \phi$. By the lemma, with D replaced by V_1, $\exists \nu_1$ such that $f_n(z)$ takes the value k at a point of V_1 for all $\nu_1 < n$, similarly $\exists \nu_2$ such that for $n > \nu_2$, $f_n(z)$ takes the value k at a point of V_2. Hence if $n > \max(\nu_1, \nu_2)$, $f_n(z) = k$ at a point both in V_1 and V_2, contradicting (iv). Hence $f(z)$ is univalent. □

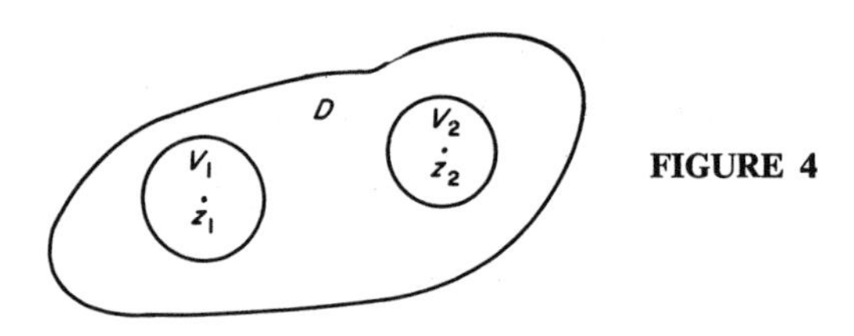

FIGURE 4

6.3.4 Definition. A set E is said to be *dense* in a set A, if $A \subseteq \bar{E}$, i.e., if every neighborhood of any point $z \in A$ contains at least one point of E.

EXAMPLE. The set E of rational points of $(0, 1)$ is dense in $A = [0, 1]$ since $A = \bar{E}$.

6.3.5 Definition. A sequence of functions $\{f_n(z)\}$ is said to be *uniformly bounded* in a set S, if $\exists M$ greater than zero such that $|f_n(z)| < M$ for $n = 1, 2, \ldots$ and all $z \in S$.

6.3.5.1 Theorem. If (i) $\{f_n(z)\}$ is a sequence of functions analytic in a domain D, (ii) $\{f_n(z)\}$ is uniformly bounded in every compact region $R \subset D$, then $\{f_n'(z)\}$ is uniformly bounded in every such R.

Proof. By the Heine–Borel theorem, it is sufficient to prove the result for a compact circular region $S \subset D$ (Fig. 5). Suppose $a \in D$ and

$$0 < \delta < \varrho(a, bd(D)).$$

Consider the compact region S: $\{z \mid |z - a| \leq \delta\}$, then $S \subset D$. Take r such that $\delta < r < \varrho(a, bd(D))$. Let C be the circle $|z - a| = r$ and write

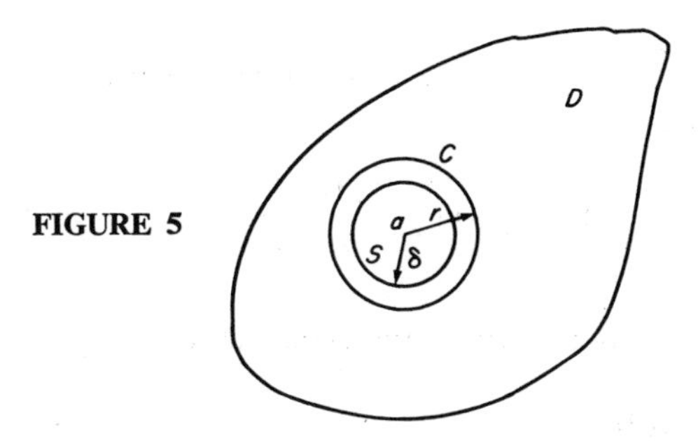

FIGURE 5

$S_r = \{z \mid |z - a| \leq r\}$, then S_r is a compact subset of D. Since $\{f_n(z)\}$ is uniformly bounded in S_r, $\exists M$ such that $|f_n(z)| < M$ for $n = 1, 2, \ldots,$ and $\forall z, z \in S_r$. Let ξ be any point of S. Then

$$f_n'(\xi) = \frac{1}{2\pi i} \int_{C+} \frac{f_n(z)}{(z - \xi)^2}\, dz.$$

For $z \in C$,

$$\left| \frac{f_n(z)}{(z - \xi)^2} \right| \leq \frac{M}{(r - \delta)^2}$$

since the distance $|z - \xi| \geq |r - \delta|$. Hence

$$|f_n'(\xi)| \leq M \cdot r/(r - \delta)^2 \quad \text{for all } \xi \in S$$

and $\{f_n'(z)\}$ is uniformly bounded in S. □

6.4 Vitali's Convergence Theorem. Let $\{f_n(z)\}$ be a sequence of functions analytic in a region D. Let $|f_n(z)| \leq M$ for all n, $z \in D$. Let $f_n(z)$ tend to a limit as $n \to \infty$, at a set of points having a limit point (i.e., point of accumulation) in D. Then $f_n(z)$ tends uniformly to a limit in a region bounded by a contour interior to D, the limit being an analytic function of z.

Proof. It is sufficient to take D a circle with limit point the center. Thus in the general case uniform convergence is proved in a circle with center the limit point interior to D. Repeating the process with any point of this circle, by analytic continuation, the domain of uniform convergence is extended to any region bounded by a contour in D. We may take the limit point to be the origin. Let the radius of the circle D, be R and let

$$f_n(z) = a_{0,n} + a_{1,n}z + \cdots.$$

Thus

$$|f_n(z) - f_n(0)| \leq |f_n(z)| + |f_n(0)| \leq 2M.$$

Also, $f_n(z) - f_n(0)$ is zero at $z = 0$. Further, since $f_n(z)$ is analytic, by Schwarz's lemma

$$|f_n(z) - f_n(0)| \leq \frac{2M}{R}|z|, \qquad |z| \leq R.$$

Let z' ($\neq 0$) be a point where the sequence converges.

$$\begin{aligned}
&|f_n(0) - f_{n+m}(0)| \\
&\quad \leq |f_n(0) - f_n(z')| + |f_n(z') - f_{n+m}(z')| \\
&\qquad + |f_{n+m}(z') - f_{n+m}(0)| \leq \frac{4M|z'|}{R} + |f_n(z') - f_{n+m}(z')|.
\end{aligned}$$

Then z' may be chosen so that the first term is arbitrarily small (by hypothesis). Also, since $f_n(z')$ approaches a limit, choosing n large the second term is arbitrarily small for all positive m. Hence $f_n(0)$, i.e., $a_{0,n}$ tends to a limit, say a_0. Consider

$$g_n(z) = [f_n(z) - a_{0,n}]/z = a_{1,n} + a_{2,n}z + \cdots.$$

This tends to a limit at z' since $a_{0,n}$ tends to a limit. For $|z| = R$, $|g_n(z)| \leq 2M/R$ [since $a_{0,n} = f_n(0)$]. Thus it is also true for $|z| < R$ (maximum modulus theorem). Consequently $g_n(z)$ satisfies the same conditions as $f_n(z)$ and $a_{1,n} \to a_1$. Similarly $a_{\nu,n} \to a_\nu$, $\forall \nu$. The convergence is uniform (with respect to ν and z), since for $|z| \leq R - \varepsilon$, by Cauchy's inequalities $|a_{\nu,n}| \leq M/R^\nu$ and since every term tends to a limit, the sum tends to a limit uniformly (Weierstrass M-test), for $|z| \leq R - \varepsilon$. □

6.5 Montel's Theorem. Let $f(z)$ be a function analytic in the half-strip S: $a < x < b$, $y > 0$. If $f(z)$ is bounded in S and $f(z) \to l$ as $y \to \infty$ for a certain fixed value ξ of x, $a < \xi < b$, then $f(z) \to l$ uniformly on every line $x = x_0$ in S. Further, $f(z) \to l$ uniformly for $a + \delta \leq x \leq b - \delta$, $\delta > 0$.

Proof. Consider a sequence of functions $f_n(z) = f(z + in)$, $n = 0, 1, \ldots,$ in the rectangle R: $a < x < b$, $0 < y < 2$. Then $f_n(z) \to l$ at every point of the line $x = \xi$ (by hypothesis), since as $y \to \infty$, $f(z)$ behaves like $f(z + in)$, $n \to \infty$. By Vitali's theorem, $f_n(z) \to l$ uniformly in a region interior to R, e.g., $a + \delta < x < b - \delta$, $\frac{1}{2} \leq y \leq \frac{3}{2}$. This is because (i) $f_n(z)$ is uniformly bounded in R by hypothesis, (ii) $f_n(z)$ is analytic for all n. $\{f(z + in)$ is a change of origin$\}$ and (iii) $x = \xi$ is a set of points each of which is a limit point of points of convergence in R. Thus $f_n(z) \to l$ uniformly on every line $x = x_0$ in R. □

NOTE. Using $f(z + in)$, $n = 0, 1, \ldots,$ means that whereas the function is analytic in the rectangle $0 < y < 2$, $a \leq x \leq b$, the whole strip $y \to \infty$ becomes the region of analyticity. Then $f(z + in^2)$ would not do, since $\exists$ gaps in the strip in which no information is forthcoming.

6.5.1 By a conformal transformation $z = i \log w$, the strip in the z-plane becomes an angle in the w-plane and the theorem states: "If $\phi(w)$ is bounded in an angle $a < \arg w < \beta$ and if $\phi(w) \to l$ as $w \to \infty$ along any line $\arg w =$ constant between α, β, then $\phi(w) \to l$ uniformly in any angle

$$\alpha + \delta \leq \arg w \leq \beta - \delta, \qquad \delta > 0.$$

CHAPTER VII

THEOREMS OF PHRAGMÉN AND LINDELÖF: MINIMUM MODULUS II

7.1 THEOREMS OF PHRAGMÉN AND LINDELÖF

These theorems are important extensions of the maximum modulus theorem and are useful in a more delicate discussion of entire functions bounded in specific directions.

A monograph of considerable extent has been written on the subject of entire functions with particular reference to results of Phragmén and Lindelöf.[†] The following theorems are a cross section of the simplest theorems contained therein.

7.2 Theorem. Let C be a simple closed contour and let $f(z)$ be analytic in and on C except at one point $P \in C$ (Fig. 1). Let $|f(z)| \leq M$ on C except at P.

Suppose $\exists$ a function $w(z)$ analytic, nonzero, and whose absolute value $|w| \leq 1$ in the region D bounded by C. Suppose further, that $w(z)$ is such that if $\varepsilon > 0$, $\exists$ a system of curves arbitrarily near to P and connecting the two sides of C around P, on which

$$|\{w(z)\}^{\varepsilon} \cdot f(z)| \leq M \qquad \text{(i.e., on } \gamma\text{, see Fig. 1)},$$

then $|f(z)| \leq M$ at all points in D.

[†] Viz. M. L. Cartwright, "Integral Functions." Cambridge Univ. Press, London and New York, 1962.

FIGURE 1

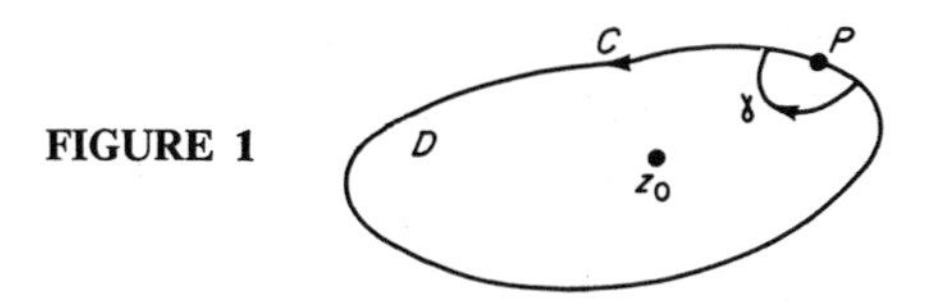

Proof. Consider the function $F(z) = \{w(z)\}^{\varepsilon} \cdot f(z)$ which is analytic in D. By hypothesis, if z_0 is in D, $\exists$ a curve surrounding z_0 on which

$$| F(z) | \leq M.$$

Thus

$$| F(z_0) | \leq M,$$

$$| f(z_0) | \leq M \, | w(z_0) |^{-\varepsilon},$$

and taking $\varepsilon \to 0$, $| f(z_0) | \leq M$. Hence the result. □

The exceptional point P may be replaced by any finite number or indeed an infinite number of points, provided that the functions $w(z)$ corresponding to them with suitable properties may be found.

Instead of starting with the previous theorem, it is usually simpler to start with a special auxiliary function adapted to the region considered. In practice, the exceptional point P is always at infinity. The previous theorem gives results about the behavior of a function in the neighborhood of an essential singularity. By a preliminary transformation the exceptional point can be placed at infinity.

7.3 Fundamental Theorem. Let $f(z)$ be an analytic function of $z = re^{i\theta}$, regular in a region D between rays making an angle π/α at the origin and on the straight lines themselves. Suppose that $| f(z) | \leq M$ on the lines and as $r \to \infty$,

$$f(z) = O(\exp r^{\beta}) \qquad \text{where } \beta < \alpha \quad \text{(uniformly).}$$

Then $| f(z) | \leq M$ throughout D.

Proof. Suppose without loss of generality that the rays are $\theta = \pm \pi/2\alpha$. Let

$$F(z) = \exp(-\varepsilon z^{\gamma}) f(z), \qquad \beta < \gamma < \alpha$$

and fixed $\varepsilon > 0$. Then

$$| F(z) | = \exp(-\varepsilon r^{\gamma} \cos \gamma\theta) \, | f(z) |. \tag{1}$$

On the lines $\theta = \pm\,\pi/2\alpha$, $\cos\gamma\theta > 0$ since $\gamma < \alpha$, thus on these lines $|\,F(z)\,| \leq |\,f(z)\,| \leq M$ (Fig. 2).

Also, on the arc $|\,\theta\,| \leq \pi/2\alpha$, $|\,z\,| = R$, and we have

$$|\,F(z)\,| \leq \exp\left(-\,\varepsilon R^{\gamma} \cos\frac{\gamma\pi}{2\alpha}\right) \cdot |\,f(z)\,|.$$

Thus

$$|\,F(z)\,| < A \exp\left(R^{\beta} - \varepsilon R^{\gamma} \cos\frac{\gamma\pi}{2\alpha}\right)$$

and the right-hand side approaches 0 as $R \to \infty$. Hence for R sufficiently large, $|\,F(z)\,| \leq M$ on this arc also. By the maximum modulus theorem, $|\,F(z)\,| \leq M$ throughout the interior of the region $|\,\theta\,| \leq \pi/2\alpha$, $r \geq R$ and since R is arbitrarily large, throughout D. Further, by (1), $|\,f(z)\,| \leq M \exp(\varepsilon r^{\gamma})$ in D and making $\varepsilon \to 0$ the result stated follows. □

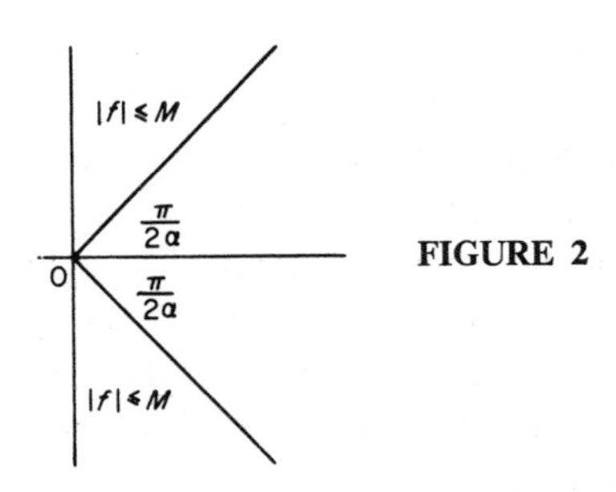

FIGURE 2

7.3.1 The straight lines may be replaced by curves approaching ∞. Notice the relationship between the angle in the theorem and the order of $f(z)$ at infinity. The wider the angle θ, the smaller the order of $f(z)$ must be for the theorem to be true (since $\beta < \alpha$).

In the next theorem, the order is not small enough for the previous proof to apply.

7.4 Theorem. The conclusion of the previous theorem still holds if $f(z) = O(\exp \delta r^{\alpha})$ for every positive δ (uniformly in the angle).

Proof. As before, take $-\,\pi/2\alpha \leq \theta \leq \pi/2\alpha$. Let $F(z) = \exp(-\,\varepsilon z^{\alpha}) f(z)$, ε fixed (temporarily). Taking $\delta < \varepsilon$ (e.g., $\varepsilon/2$), then on the real axis

$$|\,f(x)\,| = O(\exp \delta x^{\alpha})$$

and

$$|\,F(x)\,| = \exp(-\,\varepsilon x^{\alpha}) O(\exp \delta x^{\alpha}) \to 0 \qquad \text{as} \quad x \to \infty.$$

On the real axis the full benefit of the factor $\exp(-\varepsilon z^\alpha)$ is felt. Hence $|F(x)| \leq M'$ for all real x. Considering $F(z)$ in the angle $\pi/2\alpha$, write $M'' = \max(M, M')$. Then $|F(z)| \leq M''$ on the sides of the angle, and $|F(z)| = O(\exp \delta r^\alpha)$ uniformly. Since $|\exp(-\varepsilon r^\alpha \cos \theta\alpha)| < \exp \varepsilon r^\alpha$ we have $|F(z)| = O(\exp r^\beta)$ with any β such that $\alpha < \beta < 2\alpha$. Thus the previous theorem applies in each of the half-angles and gives $|F(z)| \leq M''$ in the full angle. Suppose if possible that

$$M' = \sup_{x>0} |F(x)| > M \tag{1}$$

so that $M'' = M' > M$. The supremum is not approximated as $x \to \infty$ since $F(x) \to 0$, $x \to \infty$, and $|F(0)| \leq M$ since 0 lies on the sides of the angle. Thus the supremum (1) is only approximated inside the real positive axis and is therefore attained at some point x_0. Hence $F(z)$ is analytic in the domain, $|F(z)| \leq M' = M''$ throughout the domain, yet $|F(x_0)| = M'$ for an x_0 inside. By the maximum modulus theorem, $F(z)$ is constant $= M'$ in the domain, contradicting the fact that $|F(z)| \leq M < M'$ on the sides. Hence the supposition is not possible. Therefore $M' \leq M$, $M'' = M$, and $|F(z)| \leq M$ throughout the angle. Thus $|f(z)| \leq M |\exp(\varepsilon z^\alpha)|$; now making $\varepsilon \to 0$, $|f(z)| \leq M$. □

7.5 Theorem. If $f(z) \to a$ as $z \to \infty$ along two straight lines and $f(z)$ is analytic and bounded in the angle between them, then $f(z) \to a$ uniformly in the whole angle.

Proof. We may assume that the limit a is zero. We may also assume that the angle between the two lines is less than π, since the general case may be reduced by a substitution of the form $z = w^k$. Let the lines be $\theta = \pm\theta'$, $\theta' < \pi/2$. Let

$$F(z) = \frac{z}{z+\lambda} f(z), \qquad \lambda > 0$$

and

$$|F(z)| = \frac{r}{\sqrt{r^2 + 2r\lambda\cos\theta + \lambda^2}} |f(z)| < \frac{r}{\sqrt{r^2+\lambda^2}} |f(z)|.$$

Now $|f(z)| \leq M$, say, everywhere, and

$$|f(z)| < \varepsilon \qquad \text{for } r > r_1(\varepsilon), \quad \theta = \pm\theta'.$$

Let $\lambda = r_1 M/\varepsilon$. Then

$$|F(z)| < rM/\lambda \leq \varepsilon \qquad (\text{provided } r \leq r_1)$$

and

$$|F(z)| < |f(z)| < \varepsilon \qquad \text{for} \quad r > r_1 \quad \text{and} \quad \theta = \pm\theta'.$$

Thus by the main Phragmén–Lindelöf theorem $|F(z)| \leq \varepsilon$ in the whole region. Hence

$$|f(z)| \leq \left(1 + \frac{\lambda}{r}\right) |F(z)| < 2\varepsilon \qquad \text{if} \quad r > \lambda. \quad \square$$

7.6 Theorem. If $f(z) \to a$ as $z \to \infty$ along a straight line, $f(z) \to b$ as $z \to \infty$ along another straight line, and $f(z)$ is analytic and bounded in the angle between, then (1) $a = b$ and (2) $f(z) \to a$ uniformly in the angle.

Proof. Let $f(z) \to a$ along $\theta = \alpha$ and $f(z) \to b$ along $\theta = \beta$ where $\alpha < \beta$. The function $\{f(z) - \frac{1}{2}(a+b)\}^2$ is analytic and bounded in the angle and tends to $\frac{1}{4}(a-b)^2$ on each straight line, hence it tends to this limit uniformly in the angle, i.e.,

$$\{f(z) - \tfrac{1}{2}(a+b)\}^2 - \tfrac{1}{4}(a-b)^2 = \{f(z) - a\}\{f(z) - b\}$$

tends uniformly to zero. Thus to any ε corresponds an arc on which

$$|\{f(z) - a\}\{f(z) - b\}| \leq \varepsilon.$$

At every point of this arc, either $|f(z) - a| \leq \sqrt{\varepsilon}$ or $|f(z) - b| \leq \sqrt{\varepsilon}$ (or both). We may suppose that the former inequality holds at $\theta = \alpha$ and the latter holds at $\theta = \beta$. Let θ_0 be the supremum of values of θ for which the former holds. Then θ_0 is a limit of points where the former holds and is either a point where the latter holds or a limit of such points. Since $f(z)$ is continuous, both inequalities hold at θ_0. Taking z to be this point,

$$|a - b| = |(f(z) - a) - (f(z) - b)| \leq |f(z) - a| + |f(z) - b| \leq 2\sqrt{\varepsilon}$$

Making $\varepsilon \to 0$, it follows that $a = b$, thus by the previous theorem $f(z) \to a$ uniformly. $\square$

7.7 THE PHRAGMÉN–LINDELÖF THEOREM FOR OTHER REGIONS

The angle of the previous theorem may be transformed into other regions, e.g., into a strip. Consider the fundamental theorem applied to a region $r \geq 1$, $|\theta| \leq \pi/2\alpha$. Put

$$s = i \log z, \qquad f(z) = \phi(s), \qquad s = \sigma + it.$$

Lines $\arg z = \pm \pi/2\alpha$ approach parallel lines $\sigma = \pm \pi/2\alpha$ and $t = \log |z|$ since $-i\sigma + t = \log |z| + i\theta$. Hence if $|\phi(s)| \leq M$ on the upper half of the two parallel lines and on the segment joining them (real axis), while $\phi(\sigma + it) = O(\exp e^{\varrho t})$, $\varrho < \alpha$ (since $|z| = e^t$), in the strip between them, then $|\phi(s)| \leq M$ throughout the strip.

7.8 THE PHRAGMÉN–LINDELÖF FUNCTION

Previous theorems have considered the way in which a function behaves as $z \to \infty$ in different directions. A more systematic study of this question is now made. Consider

$$f(z) = \exp\bigl((a + ib)z^{\varrho}\bigr)$$

and we have

$$|f(z)| = \exp\bigl(r^{\varrho}(a \cos \varrho\theta - b \sin \varrho\theta)\bigr).$$

The behavior of $\log |f(z)|$ depends firstly on r^{ϱ} which is independent of θ. Behavior in different directions is determined by a factor

$$h(\theta) = a \cos \varrho\theta - b \sin \varrho\theta = \log |f(z)|/r^{\varrho}.$$

Although this is a special case (Titchmarsh points out), the general case is not so different from it as may be expected. Suppose that $f(z)$ is analytic for $\alpha < \theta < \beta$ and $|z| \geq r_0$, also that $f(z)$ is of order ϱ in the angle, i.e.,

$$\overline{\lim_{r\to\infty}} \frac{\log |f(re^{i\theta})|}{r^{\varrho+\varepsilon}} = 0$$

uniformly in θ for all positive ε but not for any negative ε. Define $h(\theta)$ in general as

$$h(\theta) = \overline{\lim_{r\to\infty}} \frac{\log |f(re^{i\theta})|}{V(r)}$$

where $V(r)$ depends on a function considered. One chooses $V(r)$ such that $h(\theta)$ is finite and not identically zero: we choose the simplest case $V(r) = r^{\varrho}$ [we could choose $r^{\varrho}(\log r)^{p}(\log \log r)^{q} \cdots$].†

7.8.1 Theorem. Let

$$\alpha < \theta_1 < \theta_2 < \beta \qquad \text{and} \qquad \theta_2 - \theta_1 < \pi/\varrho.$$

† See M. L. Cartwright, "Integral Functions," p. 41. Cambridge Univ. Press, London and New York, 1962.

Let

$$h(\theta_1) \le h_1 \qquad \text{and} \qquad h(\theta_2) \le h_2.$$

Let $H(\theta)$ be the function of the form $a \cos \varrho\theta + b \sin \varrho\theta$ which takes values h_1, h_2 at θ_1, θ_2. Then

$$h(\theta) \le H(\theta), \qquad \theta_1 \le \theta \le \theta_2.$$

Proof. Observe that

$$H(\theta) = \frac{h_1 \sin \varrho(\theta_2 - \theta) + h_2 \sin \varrho(\theta - \theta_1)}{\sin \varrho(\theta_2 - \theta_1)}.$$

However, this expression is not required here. Let

$$H_\delta(\theta) = \alpha_\delta \cos \varrho\theta + b_\delta \sin \varrho\theta$$

be the H-function which equals $h_1 + \delta$, $h_2 + \delta$ ($\delta > 0$) for $\theta = \theta_1$, $\theta = \theta_2$, respectively. Let

$$F(z) = f(z) \exp(-(a_\delta - ib_\delta)z^\varrho)$$

then

$$|F(z)| = |f(z)| \exp(-H_\delta(\theta)r^\varrho),$$

thus for r sufficiently large,

$$|F(re^{i\theta_1})| = O\{\exp((h_1 + \delta)r^\varrho - H_\delta(\theta_1)r^\varrho)\} = O(1).$$

NOTE. $|f(z)| = O\{\exp((h_1 + \delta)r^\varrho\}$ since

$$h(\theta) = \overline{\lim_{r\to\infty}} \frac{\log|f(re^{i\theta})|}{r^\varrho}$$

thus $h(\theta)$ is the best (least) number such that

$$|f(re^{i\theta})| \le \exp(r^\varrho\{h(\theta) + \delta\}) \quad \text{for every } \delta > 0 \text{ and } \forall r > r_0(\delta, \theta).$$

Thus

$$|f(re^{i\theta_1})| = O(\exp(r^\varrho(h_1 + \delta)))$$

since $h(\theta_1) \le h_1$. A similar result holds for $F(re^{i\theta_2})$ and by the fundamental theorem $F(z)$ is bounded in the angle (θ_1, θ_2) uniformly.

NOTE. In order to conclude $F(z) = O(1)$ in the angle, we need $|F(z)| = O(\exp r^\beta)$ in the angle for some $\beta < \pi/(\theta_2 - \theta_1)$. Now $\pi/(\theta_2 - \theta_1) > \varrho$

by hypothesis, putting $\pi/(\theta_2 - \theta_1) = \varrho + p$ we obtain $p > 0$ and we need $|F(z)| = O(\exp r^\beta)$ for some $\beta < \varrho + p$. By hypothesis

$$|f(z)| = O(\exp r^{\varrho + p/3}),$$

i.e., $f(z)$ is of order ϱ and

$$|\exp(-(a_\delta - ib_\delta)z^\varrho)| = O(\exp kr^\varrho) = O(\exp r^{\varrho+p/3})$$

where $k = |a_\delta - ib_\delta| =$ constant. Hence indeed

$$|F(z)| = O(\exp r^\beta), \qquad \beta = \varrho + \frac{2p}{3}.$$

Thus $|f(z)| = O(\exp(H_\delta(\theta)r^\varrho))$ uniformly in the angle and by definition,

$$h(\theta) = \overline{\lim_{r\to\infty}} \frac{\log|f(re^{i\theta})|}{r^\varrho} \leq \overline{\lim_{r\to\infty}} \frac{H_\delta(\theta)r^\varrho + \text{(bounded)}}{r^\varrho} = H_\delta(\theta)$$

for $\theta_1 \leq \theta \leq \theta_2$. Since $H_\delta(\theta) \to H(\theta)$, $h(\theta) \leq H(\theta)$, $\theta_1 \leq \theta \leq \theta_2$. □

7.8.2 Characteristic Behavior of $h(\theta)$ and $H(\theta)$. We have a function analytic in the angle $\theta_1 \leq \arg z \leq \theta_2$. For each $\theta = \arg z$ in this range define $h(\theta)$ from f, i.e.,

$$h(\theta) = \overline{\lim_{r\to\infty}} \frac{\log|f(re^{i\theta})|}{r^\varrho}$$

so that $h(\theta)$ is the best number (least) such that $|f(re^{i\theta})| \leq \exp(r^\varrho\{h(\theta)+\varepsilon\})$ for every $\varepsilon > 0$ and for all $r > r_0(\varepsilon) = r_0(\varepsilon, \theta)$. Then $H(\theta)$ is a smooth function of the form

$$H(\theta) = a\cos\varrho\theta + b\sin\varrho\theta = c\cos(\varrho\theta + \alpha)$$

and is in fact the particular one which agrees with $h(\theta)$ at θ_1, and $\theta = \theta_2$ (Fig. 3). The previous theorem says $h(\theta) \leq H(\theta)$. The curve of $h(\theta)$, is "convex" in a plane geometry in which the "straight lines" are the graphs of such $H(\theta)$ functions.

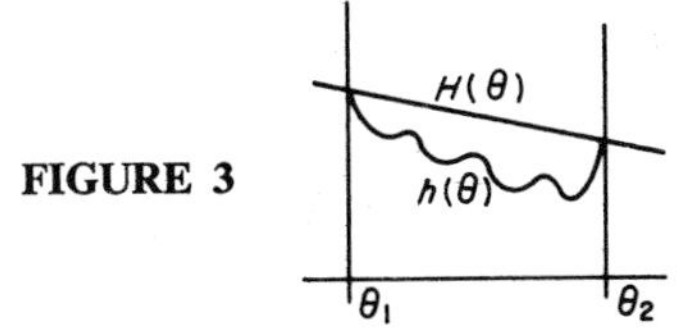

FIGURE 3

7.8.3 If

$$\overline{\lim_{r\to\infty}} \frac{\log|f(re^{i\theta})|}{r^{\varrho+\varepsilon}} = 0 \qquad \text{for all } \varepsilon > 0$$
$$\to |f| = o(\exp r^{\varrho+\varepsilon})$$
$$\to |f| = O(\exp r^{\varrho+\varepsilon})$$
$$\to \overline{\lim_{r\to\infty}} \frac{\log|f|}{r^{\varrho+\eta}} = 0 \qquad \text{for all } \eta > 0,$$

$\eta = 2\varepsilon$, for example. The first and last statements are the same, thus all are equivalent. The order ϱ is the least number for which this is true ($\exists$ a least number). For each of these statements, the numbers ϱ form a right-hand class. A second condition is needed (least or false for $\varepsilon < 0$), to make ϱ the critical value. Thus the definition for ϱ is equivalent to

$$f = O(\exp r^{\lambda}) \qquad \text{for all } \lambda > \varrho \quad \text{but} \quad \text{no } \lambda < \varrho.$$

7.8.4 From the previous theorem we may have that one or both of he $h(\theta_1)$, $h(\theta_2)$ approach $-\infty$. The conclusion is then that $h(\theta) = -\infty$ for $\theta_1 < \theta < \theta_2$. The same proof applies with one or both the h_1, h_2's being arbitrarily large and negative. Thus $h(\theta) = -\infty$ for some θ only if $h(\theta) = -\infty$ for all θ in the given angle, in which case $f(z) \equiv 0$.

7.8.5 Theorem. If $h(\theta_1)$ and $h(\theta_2)$ are finite for $\theta_1 \le \theta \le \theta_3$ and $\theta_3 - \theta_1 < \pi/\varrho$ and if $H(\theta)$ is the sinusoid such that $h(\theta_1) \le H(\theta_1)$ with $h(\theta_2) \ge H(\theta_2)$, then $h(\theta_3) \ge H(\theta_3)$.

Proof. First, $h(\theta_3) > -\infty$ by the last statement in 7.8.4. If $\exists$ a positive δ such that $h(\theta_3) \le H(\theta_3) - \delta$, consider

$$H_\delta(\theta) = H(\theta) - \delta \sin(\theta - \theta_1)\csc(\theta_3 - \theta_1).$$

Then

$$H_\delta(\theta_1) = H(\theta_1), \qquad H_\delta(\theta_2) < H(\theta_2), \qquad H_\delta(\theta_3) = H(\theta_3) - \delta$$

and thus

$$h(\theta_1) \le H_\delta(\theta_1), \qquad h(\theta_3) \le H_\delta(\theta_3)$$

and by the previous theorem,

$$h(\theta_2) \le H_\delta(\theta_2) \le H(\theta_2)$$

which is a contradiction. □

7.8.6 Theorem. If for $\theta_1 < \theta_2 < \theta_3$, $\theta_2 - \theta_1 < \pi/\varrho$, $\theta_3 - \theta_2 < \pi/\varrho$, then

$$h(\theta_1) \sin \varrho(\theta_3 - \theta_2) + h(\theta_2) \sin \varrho(\theta_1 - \theta_3) + h(\theta_3) \sin \varrho(\theta_2 - \theta_1) \geq 0.$$

Proof. For any $H(\theta)$,

$$H(\theta_1) \sin \varrho(\theta_3 - \theta_2) + H(\theta_2) \sin \varrho(\theta_1 - \theta_3) + H(\theta_3) \sin \varrho(\theta_2 - \theta_1) = 0 \tag{1}$$

Choosing $H(\theta)$ such that $H(\theta_1) = h(\theta_1)$ and $H(\theta_2) = h(\theta_2)$ then by the previous theorem $h(\theta_3) \geq H(\theta_3)$. Substitution in (1) gives the result. □

7.8.7 Theorem. The function $h(\theta)$ is continuous in any interval where it is finite.

Proof. Let $h(\theta)$ be finite in $\theta_1 \leq \theta \leq \theta_3$ and let $\theta_1 < \theta_2 < \theta_3$. Let $H_{1,2}(\theta)$ be the sinusoid which takes the values $h(\theta_1)$, $h(\theta_2)$ at θ_1, θ_2; define $H_{2,3}(\theta)$ similarly. Then by theorems 7.8.1; 7.8.5, $h(\theta) \leq H_{1,2}(\theta)$, $\theta_1 \leq \theta \leq \theta_2$, and $h(\theta) \geq H_{2,3}(\theta)$, since $h(\theta) = H_{2,3}(\theta)$ at $\theta = \theta_2, \theta_3$. Thus outside this range, in particular, $\theta_1 \leq \theta \leq \theta_2$, $h(\theta)$ exceeds or equals $H_{2,3}(\theta)$ by a previous theorem. Thus

$$H_{2,3}(\theta) \leq h(\theta) \leq H_{1,2}(\theta), \qquad \theta_1 \leq \theta \leq \theta_2$$

and similarly,

$$H_{1,2}(\theta) \leq h(\theta) \leq H_{2,3}(\theta), \qquad \theta_2 \leq \theta \leq \theta_3.$$

Hence in whichever of these intervals θ lies

$$\frac{H_{1,2}(\theta) - H_{1,2}(\theta_2)}{\theta - \theta_2} \leq \frac{h(\theta) - h(\theta_2)}{\theta - \theta_2} \leq \frac{H_{2,3}(\theta) - H_{2,3}(\theta_2)}{\theta - \theta_2}.$$

The extreme terms tend to limits as $\theta \to \theta_2$ [the same since $H_{1,2}(\theta_2) = H_{2,3}(\theta_2)$] and $|h(\theta) - h(\theta_2)| < K|\theta - \theta_2|$ for k independent of θ. Hence $h(\theta)$ is continuous at θ_2 and continuous everywhere in the interval. Also it has a left-hand and right-hand derivative at θ_2. □

7.9 MINIMUM MODULUS

We now study $m(r)$ in more detail. We show in particular that for functions of small order not only can $m(r)$ be large, but for a large proportion of r.

7.9.1 Theorem. If $\varrho < \frac{1}{2}$, there is a sequence of values of r tending to infinity through which $m(r) \to \infty$.

Proof. There is no line $\arg z =$ constant on which $f(z)$ is bounded, since the whole plane bounded by this line forms an angle $2\pi < \pi/\varrho$ if $\varrho < \frac{1}{2}$. Hence if $f(z)$ is bounded on this line, it is bounded everywhere and so reduces to a constant. Let

$$f(z) = cz^k \prod_{(n)} \left(1 - \frac{z}{z_n}\right)$$

and

$$\phi(z) = cz^k \prod_{(n)} \left(1 + \frac{z}{r_n}\right), \qquad r_n = | z_n |$$

(i.e., ϕ has real negative zeros). Then since $| 1 - (z/z_n) | \geq | 1 - (r/r_n) |$,

$$m_f(r) \geq m_\phi(r) = | \phi(-r) | \qquad \text{and} \qquad M_f(r) \leq \phi(r) = M_\phi(r).$$

Thus $\min_{|z|=r} | f(z) | \geq | \phi(-r) |$. Also $\phi(-r)$ is unbounded since $\phi(z)$ is an entire function of the same order as $f(z)$. Then $\phi(-r)$ is unbounded in particular along the negative real axis as $r \to \infty$. Hence it is possible to find values of r, tending to ∞ along which $m(r)$ is unbounded. □

We prove a lemma now which enables us to establish a far more precise result.

7.9.2 Lemma. Given that

$$f(z) = \prod_{n=1}^{\infty} \left(1 + \frac{z}{r_n}\right), \qquad r_n > 0$$

and f is of order ϱ, $0 < \varrho < 1$, then for $\varrho < \sigma < 1$,

$$\int_0^\infty \frac{\log f(x)}{x^{\sigma+1}}\, dx = \frac{\pi}{\sigma \sin \pi\sigma} \sum_{n=1}^{\infty} \frac{1}{r_n{}^\sigma}$$

and

$$\int_0^\infty \frac{\log | f(-x) |}{x^{\sigma+1}}\, dx = \frac{\pi}{\sigma \tan \pi\sigma} \sum_{n=1}^{\infty} \frac{1}{r_n{}^\sigma}.$$

Proof.

$$\int_0^\infty \frac{\log f(x)}{x^{s+1}}\, dx = \int_0^\infty \left\{ \sum_{(n)} \frac{\log(1 + (x/r_n))}{x^{s+1}} \right\} dx$$

$$= \sum_{(n)} \int_0^\infty x^{-1-s} \log\left(1 + \frac{x}{r_n}\right) dx \tag{1}$$

provided

$$\sum_{(n)} \int_0^\infty \left| x^{-1-s} \log\left(1 + \frac{x}{r_n}\right) \right| dx < \infty. \tag{2}$$

The sum in (2) is, on writing $\sigma = \mathscr{R}(s)$,

$$\sum_{(n)} \int_0^\infty x^{-1-\sigma} \log\left(1 + \frac{x}{r_n}\right) dx = \sum_{(n)} r_n^{-\sigma} \int_0^\infty t^{-1-\sigma} \log(1 + t)\, dt$$

which is less than ∞ if $\sigma > \varrho_1$, in particular for $0 < \varrho < \sigma < 1$. On that condition,

$$\int_0^\infty \frac{\log f(x)}{x^{s+1}} dx = \sum_{(n)} r_n^{-s} \int_0^\infty \frac{\log(1 + t)}{t^{1+s}} dt = \frac{\pi}{s \sin \pi s} \sum_{(n)} r_n^{-s}.$$

To establish this, we observe that

$$\int_0^\infty \frac{\log(1 + t)}{t^{1+s}} dt = \left[-\frac{\log(1 + t)}{st^s} \right]_0^\infty + \int_0^\infty \frac{dt}{st^s(1 + t)}$$

$$= \frac{1}{s} \int_0^\infty \frac{t^{-s}}{1 + t} dt = \frac{\pi}{s \sin \pi s}$$

by residues, or

$$\int_0^1 \frac{\log(1 + t)}{t^{1+s}} dt = \int_0^1 t^{-1-s}\left(t - \frac{t^2}{2} + \cdots\right) dt$$

$$= \frac{1}{1 - s} - \frac{1}{2(2 - s)} + \cdots,$$

$$\int_1^\infty \frac{\log(1 + t)}{t^{1+s}} dt = \int_1^\infty t^{1-s}\left\{\log t + \log\left(1 + \frac{1}{t}\right)\right\} dt$$

$$= -\frac{\log t}{st^s}\Big]_1^\infty + \int_1^\infty \frac{dt}{st^s \cdot t}$$

$$+ \int_1^\infty t^{-1-s}\left(\frac{1}{t} - \frac{1}{2t^2} + \cdots\right) dt$$

$$= 0 + \frac{1}{s^2} + \frac{1}{1 + s} - \frac{1}{2(2 + s)} + \cdots.$$

Combining the two results we obtain

$$\int_0^\infty \frac{\log(1 + t)}{t^{1+s}} dt = \frac{1}{s^2} - \sum_{n=1}^\infty \left\{ \frac{(-1)^n}{n(n - s)} + \frac{(-1)^n}{n(n + s)} \right\}$$

$$= \frac{1}{s^2} - \sum_{n=1}^\infty (-1)^n \cdot \frac{2}{n^2 - s^2} = \frac{\pi}{s \sin \pi s}.$$

Similarly

$$\begin{aligned}
\int_0^\infty \frac{\log|f(-x)|}{x^{s+1}}\,dx &= \int_0^\infty x^{-s-1} \sum_{(n)} \log\left|1-\frac{x}{r_n}\right| dx \\
&= \sum_{(n)} \int_0^\infty x^{-s-1} \log\left|1-\frac{x}{r_n}\right| dx \\
&= \sum_{(n)} r_n^{-s} \int_0^\infty t^{-s-1} \log|1-t|\,dt \\
&= \frac{\pi}{s\tan \pi s} \sum_{(n)} r_n^{-s},
\end{aligned}$$

provided that

$$\sum_{(n)} \int_0^\infty \left| x^{-s-1} \log\left|1-\frac{x}{r_n}\right|\right| dx < \infty \tag{3}$$

and that

$$\int_0^\infty t^{-s-1} \log|1-t|\,dt = \frac{\pi}{s\tan \pi s}. \tag{4}$$

Condition (3) reduces to

$$\sum_{(n)} \int_0^\infty x^{-\sigma-1} \left|\log\left|1-\frac{x}{r_n}\right|\right| dx < \infty$$

or

$$\left(\sum_{(n)} r_n^{-\sigma}\right) \int_0^\infty t^{-\sigma-1} \left|\log|1-t|\right| dt < \infty.$$

The integral is finite if $0 < \sigma < 1$, while the series converges if $\sigma > \varrho_1$, i.e., $\mathscr{R}(s) > \varrho_1$ as before. To prove (4),

$$\int_0^1 t^{-s-1} \log|1-t|\,dt = \sum_{n=1}^{\infty} \frac{-1}{n(n-s)}, \qquad |1-t| = 1-t,\ 0 \le t \le 1,$$

$$\int_1^\infty t^{-s-1} \log|1-t|\,dt = \frac{1}{s^2} - \sum_{n=1}^{\infty} \frac{1}{n(n+s)},$$

$$|1-t| = t-1 = t(1-1/t), \quad t > 1.$$

and

$$\int_0^\infty t^{-s-1} \log |1 - t| \, dt = \frac{1}{s^2} - \sum_{n=1}^{\infty} \frac{2}{n^2 - s^2}$$
$$= \frac{1}{s}\left\{\frac{1}{s} + \sum_{n=1}^{\infty} \frac{2s}{s^2 - n^2}\right\}$$
$$= \frac{\pi}{s \tan \pi s}. \quad \square$$

The more precise result now is the following theorem.

7.9.3 Theorem (Pólya). If $0 < \varrho < 1$, there are arbitrarily large values of r for which $m(r) > \{M(r)\}^{\cos \pi\varrho - \varepsilon}$.

Proof. Defining $f(z)$ and $\phi(z)$ as in theorem 7.9.1, we take $c = 1$, $k = 0$. It is sufficient to prove the theorem for $\phi(z)$. If z' is a point where $|f(z')| = m(r)$, then

$$|\phi(r)\phi(-r)| = \left| \prod_{n=1}^{\infty} \left(1 - \frac{r^2}{r_n^{\,2}}\right)\right| \leq |f(z')f(-z')| \leq m(r)M(r).$$

Thus if the theorem is true for $\phi(z)$, since $|\phi(-r)| > \{\phi(r)\}^{\cos\pi\varrho-\varepsilon}$ for arbitrarily large r (by hypothesis), then

$$m(r)M(r) \geq |\phi(r)|^{1+\cos\pi\varrho-\varepsilon} \geq \{M(r)\}^{1+\cos\pi\varrho-\varepsilon}.$$

Thus for $f(z)$, $m(r) > \{M(r)\}^{\cos\pi\varrho-\varepsilon}$, $r \gg 1$.

We now establish the theorem for $\phi(z)$. If the theorem is false for $\phi(z)$, $\exists$ positive constants ε, a such that $a \geq 1$, and

$$\log |\phi(-x)| < (\cos \pi\varrho - \varepsilon) \log \phi(x), \qquad \forall x > a.$$

For $0 < \varrho < \mathscr{R}(s) < 1$ by the previous lemma,

$$\int_0^\infty \{\cos \pi s \log \phi(x) - \log |\phi(-x)|\} x^{-s-1} \, dx = 0.$$

Thus

$$F(s) = \int_a^\infty \{\cos \pi s \log \phi(x) - \log |\phi(-x)|\} x^{-s-1} \, dx$$
$$= -\int_0^a \{\cos \pi s \log \phi(x) - \log |\phi(-x)|\} x^{-s-1} \, dx$$

at least in $\varrho < \mathscr{R}(s) < 1$ where the last two integrals exist. Now

$$-\int_0^a \{\cos \pi s \log \phi(x) - \log | \phi(-x) |\} x^{-s-1}\, dx$$

is regular at every point in $[0, a]$ except zero. At zero,

$$\phi(x) = \phi(0) + x\phi'(0) + \cdots.$$

Thus

$$\{\cos \pi s \log \phi(x) - \log | \phi(-x) |\} x^{-s-1} \sim A(\cos \pi s + 1) x^{-s}$$

where $\log \phi(x) \sim Ax$ and $A = \phi'(0) = \sum_{(n)} 1/r_n$. Hence for x in the neighborhood of zero, $\int_0$ converges only if $\mathscr{R}(s) < 1$. [The condition $\varrho < \mathscr{R}(s)$ is required for convergence of the integral at ∞.] Consequently $-\int_0^a \{\cdots\} x^{-s-1}\, dx$ converges and defines a function of s regular in $0 < \mathscr{R}(s) < 1$. We know that $F(s)$, which is given in a narrower strip $\varrho < \mathscr{R}(s) < 1$ by $\int_a^\infty$, is in fact regular (i.e., analytically continuable) in the wider strip $0 < \mathscr{R}(s) < 1$. Taking

$$\begin{aligned}
\phi_1(x) &= (\cos \pi\varrho - \varepsilon) \log \phi(x) - \log | \phi(-x) |,\\
\phi_2(x) &= \log \phi(x),\\
\psi(s) &= \cos \pi s - \cos \pi\varrho + \varepsilon,
\end{aligned}$$

then

$$F(s) = \int_a^\infty \{\phi_1(x) + \psi(s)\phi_2(x)\} x^{-s-1}\, dx$$

and $\phi_1(x) > 0$, $\forall x > a$ (by hypothesis), also, $\phi_2(x) > 0$. Figure 4 graphs $\psi(s)$ (s real).

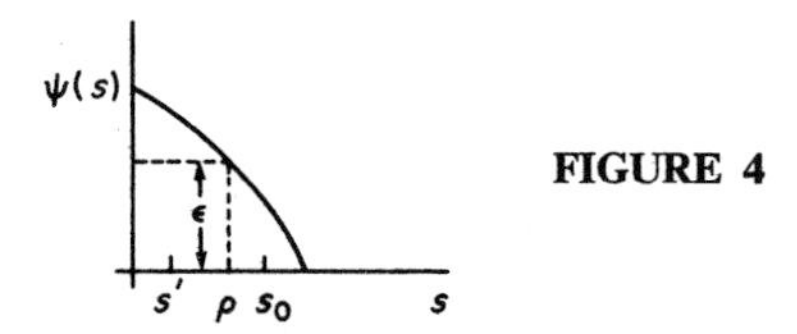

FIGURE 4

NOTE. We use s complex since we must know that $F(s)$ has a circular disk of regularity with center $s_0 > \varrho$ and enclosing $s' = s_0 - h$. We use in effect the Taylor series about s_0, with displacement $-h$. In (I) of Fig. 5,

$$F(s) = \int_a^\infty = -\int_0^a.$$

In (I) and (II), $F(s)$ is regular and is $-\int_0^a$ only.

FIGURE 5

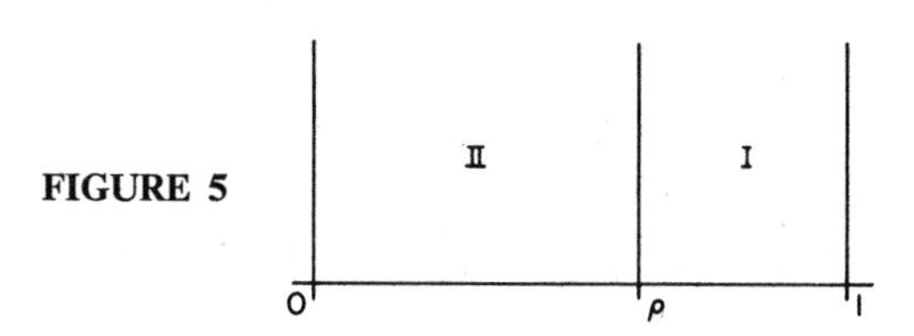

The underlying idea now, is that since $F(s)$ defines a regular function for s real and greater than ϱ and since $F(s)$ is in fact regular right down to $s = 0$, it ought to hold also at points $s = s' < \varrho$, particularly as in the range of s being considered, $\psi(s)$ is positive so that there can be no cancellation of positive and negative contributions in $(\phi_1 + \psi \cdot \phi_2)x^{-s-1}$. If $F(s)$ were regular at s', i.e.,

$$F(s') = \int_a^\infty \{\phi_1(x) + \psi(s')\phi_2(x)\}x^{-s'-1}\,dx$$

we should obtain

$$\varepsilon \int_a^\infty \phi_2(x)x^{-s'-1}\,dx < \psi(s') \int_a^\infty \phi_2(x)x^{-s'-1}\,dx$$

since then $\cos \pi s' > \cos \pi\varrho$ (s' real) and

$$\begin{aligned}\varepsilon \int_a^\infty \phi_2(x)x^{-s'-1}\,dx &< \int_a^\infty \{\phi_1(x) + \psi(s')\phi_2(x)\}x^{-s'-1}\,dx \\ &= F(s') < \infty,\end{aligned}$$

whence $\sum_{(n)} r_n^{-s'} < \infty$ contrary to the definition of ϱ. However, we cannot simply apply $\int_a^\infty \{\phi_1(x) + \psi(s)\phi_2(x)\}x^{-s-1}\,dx$ at s'. We have a function $F(s)$ defined in the whole strip $0 < \mathscr{R}(s) < 1$ but we only know that it satisfies $\int_a^\infty \{\phi_1 + \psi\phi_2\}x^{-s-1}\,dx = F(s)$ for s in a narrower strip $\varrho < \mathscr{R}(s) < 1$. Choose s_0 real and greater than ϱ but so near to ϱ that $\exists$ a circle centered at s_0 with ϱ inside it and which lies entirely in the strip $0 < \mathscr{R}(s) < 1$ in which we know that $F(s)$ is regular (Fig. 6). Further, we suppose that

$$0 < s_0 - \varrho < h < k < 1 < \min(s_0, 1 - s_0).$$

On the circle centered at s_0 and radius l, $F(s)$ is regular and therefore bounded, say $|F(s)| \le A$ on $|s - s_0| = l$. Hence

$$|F^{(n)}(s_0)/n!| \le A/l^n \qquad \text{(Cauchy's inequality)}.$$

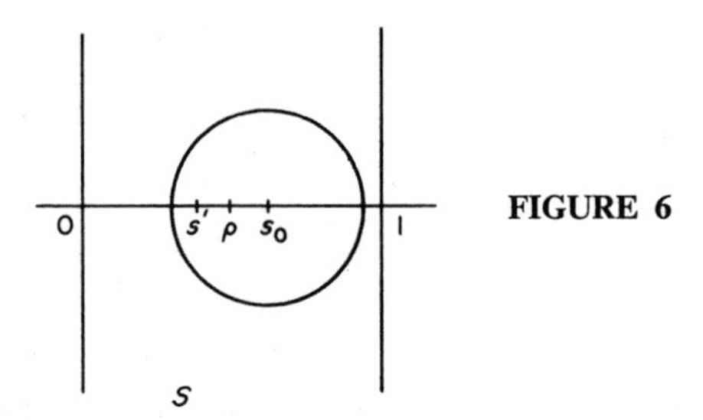

FIGURE 6

Thus with $s' = s_0 - h$ $(s' < \varrho)$, $D \equiv d/ds$, $h > 0$, we have

$$F(s') = e^{-hD}F(s_0) \qquad \text{(Taylor expansion about } s_0)$$
$$= \sum_{n=0}^{\infty} \frac{(-h)^n}{n!} F^{(n)}(s_0) \equiv \sum_{n=0}^{\infty} U_n,$$

where $|U_n| \le M_n = h^n A/l^n$. Then

$$\sum_{(n)} M_n = \frac{A}{1-(h/l)} = B,$$

and $\sum_{(n)} U_n$ converges uniformly. Note that

$$\left(1 - \frac{hD}{m}\right)^m F(s_0) = \sum_{n=0}^{m} \frac{(-h)^n}{n!} F^{(n)}(s_0) \prod_{\mu=1}^{n-1}\left(1 - \frac{\mu}{m}\right)$$
$$= \sum_{n=0}^{m} U_{n,m}$$

where $|U_{n,m}| \le |U_n| \le M_n$. Thus we have that the series $\sum_{(n)} U_n$ converges to $F(s')$ and

$$\sum_{(n)} U_{n,m} \to \sum_{(n)} U_n = F(s') \qquad \text{as} \quad m \to \infty.$$

There is no difficulty in principle in evaluating

$$D^{(n)}F(s_0) = F^{(n)}(s_0)$$

by differentiating $\int_a^\infty \{\phi_1 + \psi\phi_2\}x^{-s-1}\,dx$ under the sign of integration. Thus we have

$$F(s') = \sum_{n=0}^{\infty} \frac{(-h)^n}{n!} \left[\int_a^\infty \left(\frac{d}{ds}\right)^n \{\phi_1 + \psi\phi_2\}x^{-s-1}\,dx\right]_{s=s_0}.$$

The difficulty arises when we try to justify taking the summation under the sign of integration. To meet this difficulty, Pólya (whose method we are following essentially) considers the finite object $(1 - (hD/m))^m$ instead of e^{-hD}. As $m \to \infty$, the operator $(1 - (hD/m))^m$ tends to e^{-hD} and by the above analysis $(1 - (hD/m))^m F(s_0) \to F(s')$ as $m \to \infty$. For each m, $\sum_{(n)} U_{n,m}$ is finite and therefore the summation is valid under the sign of integration. We now consider $(1 - (hD/m))^m F(s_0)$.

$$\text{(i)}\quad \left(1 - \frac{hD}{m}\right)^m \phi_1(x) x^{-s-1} = x^{-s-1}\phi_1(x)\left(1 + \frac{h\xi}{m}\right)^m.$$

[Transform $x = e^{\xi}$ and consider $(1 + (h\xi/m) - (hD/m))^m \cdot 1$.]

$$\text{(ii)}\quad \left(1 - \frac{hD}{m}\right)^m \psi(s)\phi_2(x) x^{-s-1} = x^{-s-1}\phi_2(x)\left(1 + \frac{h\xi}{m} - \frac{hD}{m}\right)^m \psi(s)$$

$$= x^{-s-1}\phi_2(x)\left(1 + \frac{h\xi}{m}\right)^m \{\psi(s) + E\}$$

where

$$E = \sum_{\mu=1}^{m} \left(1 + \frac{h\xi}{m}\right)^{-\mu} \left(-\frac{h}{m}\right)^{\mu} \psi^{(\mu)}(s) \binom{m}{\mu}.$$

It will be sufficient for our purpose to establish that at $s = s_0$,

$$E \geq -\tfrac{1}{2}\psi(s), \qquad \forall m, \xi.$$

We have

$$|E| \leq \sum_{\mu=1}^{m} \left(\frac{h}{m}\right)^{\mu} |\psi^{(\mu)}(s)| \frac{m(m-1)\cdots(m-\mu+1)}{\mu!}.$$

Now $|\psi^{(\mu)}(s)| \leq \pi^{\mu}$ and ξ must be positive for we require

$$\left(1 + \frac{h\xi}{m}\right)^{m-\mu} < \left(1 + \frac{h\xi}{m}\right)^m.$$

Thus

$$|E| \leq \sum_{\mu=1}^{m} \frac{(\pi h)^{\mu}}{\mu!} < e^{\pi h} - 1$$
$$\leq 2h(e^{\pi/2} - 1)$$

since h is certainly less than $\frac{1}{2}$. Consequently $|E| < 9h$ and $E \geq -\frac{1}{2}\psi(s)$ provided

$$9h \leq \tfrac{1}{2}\psi(s_0). \tag{1}$$

Now, by hypothesis, $\psi(\varrho) = \varepsilon$ is positive and (1) shows that we want s_0

so near to ϱ that ψ is still positive there. Moreover we require $s_0 - \varrho < h$ (i.e., $s' < \varrho$). Thus we require $18(s_0 - \varrho) < \psi(s_0)$. This is true at $s_0 = \varrho$ and by continuity of ψ, for all s_0 sufficiently near ϱ (1) can be satisfied. The case that concerns us is $s_0 > \varrho$ whence

$$\psi(s_0) > \psi(\varrho) - \pi(s_0 - \varrho)$$

since

$$\begin{aligned}\psi(\varrho) - \psi(s_0) &= \varepsilon - \cos \pi s_0 + \cos \pi\varrho - \varepsilon \\ &= 2 \sin \frac{\pi}{2}(s_0 + \varrho) \cdot \sin \frac{\pi}{2}(s_0 - \varrho) \\ &\leq 2 \sin \frac{\pi}{2}(s_0 - \varrho) < \pi(s_0 - \varrho)\end{aligned}$$

because $s_0 - \varrho$ is very small. Thus it is sufficient to choose s_0 so that

$$(18 + \pi)(s_0 - \varrho) < \varepsilon$$

(there are other conditions such as $s_0 - \varrho < 1 - s_0$). Now we can indeed arrange that $E \geq -\frac{1}{2}\psi(s)$ and hence

$$\begin{aligned}\left(1 - \frac{hD}{m}\right)^m F(s_0) &\geq \int_a^\infty x^{-s_0-1}\left(1 + \frac{h\xi}{m}\right)^m \left\{\phi_1(x) + \frac{1}{2}\phi_2(x)\psi(s_0)\right\} dx \\ &\geq \int_a^\infty x^{-s_0-1}\left(1 + \frac{h\xi}{m}\right)^m \frac{1}{2}\phi_2(x)\psi(s_0)\, dx.\end{aligned}$$

As $m \to \infty$, everything on the right-hand side is positive and increases while the left-hand side is bounded by B, in fact it tends to $F(s_0 - h) = F(s')$. Hence

$$B \geq \int_a^\infty x^{-s_0-1} e^{h\xi} \tfrac{1}{2}\, \phi_2(x)\psi(s_0)\, dx$$

and since $\xi = \log x > 0$

$$\int_a^\infty x^{-s'-1}\phi_2(x)\, dx < 2B/\psi(s_0) = C.$$

In particular, the integral on the left-hand side converges. Hence, apparently $\sum_{(n)} r_n^{-s'}$ converges for a value of the exponent less than ϱ. This contradiction shows that the original supposition was untrue and the theorem is proved. □

For functions of order 1 and exponential type, i.e., functions $f(z) = O(e^{A|z|})$ we have the following theorem:

7.9.4 Theorem. If $f(z) = O(e^{kr})$, there are arbitrarily large values of r for which $m(r) > e^{-(k+\varepsilon)r}$.

Proof. Let

$$f(z) = ce^{az} \prod_{n=1}^{\infty} \left(1 - \frac{z}{z_n}\right) e^{z/z_n}$$

and

$$\phi(z) = c^2 \prod_{n=1}^{\infty} \left(1 + \frac{z}{r_n{}^2}\right), \qquad \text{where} \quad |z_n| = r_n.$$

Since $f(z) = O(e^{k|z|})$, $n(r) = O(r)$, thus $\exists$ constant k such that $1/r_n < k/n$ since $n(r_n) = n = O(r_n)$. Thus

$$|\phi(z)| < |c|^2 \prod_{n=1}^{\infty} \left(1 + \frac{k^2 r}{n^2}\right) = |c|^2 \frac{\sinh(\pi k \sqrt{r})}{\pi k \sqrt{r}} = O(e^{\pi k \sqrt{r}}),$$

(cf. product for $\sin\theta$). Let

$$h(\theta) = \overline{\lim_{r\to\infty}} \frac{\log|\phi(re^{i\theta})|}{\sqrt{r}}.$$

Then $h(\theta) \leq \pi k$, $\forall\theta$. Since $|\phi(z)| \geq |c|^2$ if $\mathscr{R}(z) \geq 0$, we see that $h(\theta)$ is finite for $-\pi/2 \leq \theta \leq \pi/2$. Thus by a previous theorem if $h(\theta)$ is finite somewhere, it is finite everywhere. Also $h(-\theta) = h(\theta)$ and with $0 < \theta < \pi$

$$\theta_1 = -\theta, \qquad \theta_2 = 0, \qquad \theta_3 = \theta, \qquad \varrho = \tfrac{1}{2},$$

we have

$$h(\theta) \sin\frac{\theta}{2} - h(0) \sin\theta + h(\theta) \sin\frac{\theta}{2} \geq 0$$

from Theorem 7.8.6 hence

$$h(\theta) \geq h(0) \cos\theta/2. \tag{1}$$

Since $|\phi(z)| > |c|^2$ for z real and positive

$$h(0) \geq 0 \qquad \left(\text{for } \frac{\log|\phi(r)|}{\sqrt{r}} \to 0 \text{ or is positive}\right).$$

Therefore $h(\theta) \geq 0$ for $-\pi \leq \theta \leq \pi$ [from (1)] and since $h(\theta)$ is continuous, $h(\pi) \geq 0$. Thus

$$|\phi(-r)| = |\phi(re^{i\pi})| > \exp\left(-\varepsilon\sqrt{r}\right) \quad \text{for large } r,$$

from the definition of $h(\theta)$. For such r,

$$|f(z)f(-z)| = \left|c^2 \prod_{n=1}^{\infty}\left(1 - \frac{z^2}{z_n^2}\right)\right| \geq |\phi(-r^2)| > e^{-\varepsilon r}$$

and

$$|f(z)| > e^{-\varepsilon r}/|f(-z)| > Ae^{-(k+\varepsilon)r}$$

whence $m(r) > e^{-(k+\varepsilon)r}$. □

A considerable amount of work has been done on the minimum modulus of functions of both finite and infinite order. For example, Hayman† has shown that for functions of infinite order, $m(r) > M(r)^{-A \text{ logloglog } m(r)}$ for some arbitrarily large r provided $A > e^2/\pi$. More recently, Kjellberg‡ has proved a much more delicate result than Theorem 7.9.3. He shows that for each non constant entire function $f(z)$ and for each λ satisfying $0 < \lambda < 1$, the following holds: Either

$$(1) \quad \log m(r) > \cos \pi\lambda \log M(r)$$

for certain arbitrarily large values of r, or if (1) is not fulfilled, the limit

$$(2) \quad \lim_{r\to\infty} \log M(r)/r^{\lambda}$$

exists and is positive or is infinite. This is, of course, a considerable improvement on the similar theorem we have already proved.

† W. K. Hayman, "The minimum modulus of large integral functions," *Proc. London Math. Soc.* (3) **2** (1952) 469–512.

‡ B. Kjellberg, "A theorem on the minimum modulus of entire functions," *Math. Scand.* **12** (1963) 5–11.

CHAPTER VIII

THEOREMS OF BOREL, SCHOTTKY, PICARD, AND LANDAU: ASYMPTOTIC VALUES

8.1 THE a-POINTS OF AN ENTIRE FUNCTION AND EXCEPTIONAL VALUES

Rather than discussing the zeros of a function, we now consider the distribution of points where the function takes any value a, the a-points. The following result has already been obtained: "If $f(z)$ is of finite order ϱ where ϱ is not an integer, then $f(z)$ has an infinity of zeros and the exponent of convergence of the zeros is ϱ."

Clearly $f(z) - a$ is also of order ϱ where a is any constant. Hence $f(z)$ has an infinity of a-points and their exponent of convergence is ϱ, i.e., their density is roughly the same for all values of a.

Similarly for functions of zero order, such a function has an infinity of zeros unless it reduces to a polynomial. We can obtain similar results for functions of positive integral order by using Hadamard's factorization theorem, but this tells us nothing about functions of infinite order. We shall show, using methods which are independent of concepts of order, Picard's theorem, viz., "An entire function which is not constant takes every value with one possible exception, an infinite number of times." Picard's proof used elliptic modular functions. A further proof is now given depending upon Schottky's theorem. As Titchmarsh points out, the main feature of

Picard's theorem is that it admits the possibility of there being one exceptional value. The value may actually exist, e.g., $e^z \neq 0$. A value with this property is called *exceptional-P*. There is another sense in which a value may be exceptional. A function may take the value a but only at points which have a convergence exponent less than ϱ. For example

$$f(z) = e^z \cos\sqrt{z}$$

has $\varrho = 1$, but the zeros are

$$z_n = \{(2n + 1)\pi/2\}^2, \qquad n = 0, \ \pm 1, \ \ldots .$$

Thus the exponent of convergence is $\frac{1}{2}$. A value with this property is called *exceptional-B*, i.e., in the sense of Borel. A value which is exceptional-P is necessarily exceptional-B. For entire functions of positive integral order, Picard's theorem is a consequence of the following theorem of Borel which shows that $\exists$ not merely at most one value exceptional-P, but that $\exists$ at most one value exceptional-B. Alternative proofs of Picard's theorem can be given depending upon theorems of Bloch.†

8.2 Borel's Theorem. If the order of the entire function $f(z)$ is a positive integer, then the exponent of convergence of the a-points of $f(z)$ is equal to the order, with one possible exceptional value of a.

Proof. Suppose $\exists$ two exceptional values a and b. Then

$$f(z) - a = z^{k_1} e^{Q_1(z)} P_1(z) \tag{1}$$

and

$$f(z) - b = z^{k_2} e^{Q_2(z)} P_2(z) \tag{2}$$

where $Q_1(z)$ and $Q_2(z)$ are polynomials of degree ϱ and $P_1(z)$ and $P_2(z)$ are canonical products of order less than ϱ. Note the order of $f(z) - a$ equals ϱ which is $\max(\varrho, \varrho_1)$ and ϱ_1 the order of P_1 is assumed less than ϱ since a is an exceptional value. Subtracting,

$$b - a = z^{k_1} e^{Q_1(z)} P_1(z) - z^{k_2} e^{Q_2(z)} P_2(z) \tag{3}$$

and

$$z^{k_1} P_1(z) e^{Q_1(z) - Q_2(z)} = z^{k_2} P_2(z) + (b - a) e^{-Q_2(z)}. \tag{4}$$

† See P. Dienes, "The Taylor Series," Chapter VIII. Oxford Univ. Press, London and New York, 1931.

Since $Q_2(z)$ is of degree ϱ, the right-hand side is of order ϱ. Thus the left-hand side is of order ϱ and $Q_1(z) - Q_2(z)$ is of degree ϱ since $P_1(z)$ is of order less than ϱ. Differentiating (3) we obtain,

$$(z^{k_1}P_1Q_1' + k_1z^{k_1-1}P_1 + z^{k_1}P_1')e^{Q_1} = (z^{k_2}P_2Q_2' + k_2z^{k_2-1}P_2 + z^{k_2}P_2')e^{Q_2}.$$

The order of P_1' is the same as P_1 (see Theorem 5.6) and is less than ϱ, hence the coefficient of e^{Q_1} is of order less than ϱ and similarly, so is that of e^{Q_2}. Since the brackets are entire functions we may write them as $z^{k_3}P_3e^{Q_3}$ and $z^{k_4}P_4e^{Q_4}$, respectively, where Q_3 and Q_4 are polynomials of degree $\varrho - 1$ at most (since ϱ is an integer and the order of each bracket is less than ϱ). Also, P_3 and P_4 are canonical products, thus

$$z^{k_3}P_3e^{Q_1+Q_3} = z^{k_4}P_4e^{Q_2+Q_4},$$

the two sides having the same zeros $k_3 = k_4$ and $P_3 = P_4$. Thus

$$Q_1 + Q_3 = Q_2 + Q_4 \qquad \text{or} \qquad Q_1 - Q_2 = Q_4 - Q_3$$

and $Q_4 - Q_3$ is of degree not exceeding $\varrho - 1$, whereas $Q_1 - Q_2$ was of degree ϱ. The contradiction proves the theorem. □

8.3 We proceed with the theory of exceptional values by studying meromorphic functions rather than entire functions. The exponent of convergence of the poles of a meromorphic function $f(z)$ is defined in the same way as for zeros of an entire function. If the exponent of convergence of the poles is finite, their canonical product $H(z)$ is of finite order. We define the order ϱ of $f(z)$ as the larger of the orders of $H(z)$ and $G(z)$ where $G(z) = z^mH(z)f(z)$, m being the order of the pole at the origin (cf., Theorem 4.9.1). If the exponent of convergence of the poles is infinite, the order of $f(z)$ is infinite. If the exponent of convergence of the poles is less than the order of $f(z)$, then ∞ is an exceptional-B value. Any three exceptional values, a, b, c, can be reduced to 0, 1, ∞ by putting

$$F(z) = \frac{(f(z) - a)(b - c)}{(f(z) - c)(b - a)},$$

and the order of $F(z)$ is equal to the order of $f(z)$. If $f(z)$ is an entire function, it is usual to confine the discussion of exceptional values to finite exceptional values.

We extend Borel's theorem a little, by the following theorem.

8.4 Theorem. Meromorphic functions of finite nonintegral order have at most one exceptional-B value. In particular, entire functions of finite nonintegral order have no exceptional-B values.

Proof. Suppose that the order ϱ of $f(z)$ is not an integer and suppose that the values a and c are exceptional-B (implying that $\varrho > 0$). We may suppose that $a = 0$ and $c = \infty$ and write $f(z) = G(z)/H(z)$ where the factor due to a pole at the origin has been absorbed in $H(z)$. Since the exponents of convergence of the zeros and poles of $f(z)$ are each less than ϱ, the order of $H(z)$ is less than ϱ and the order of $G(z)$ is less than ϱ, unless $\varrho = q$. However, q is an integer and ϱ is not, and further, ϱ is the larger of the orders of G and H so we have a contradiction. Thus, either a or c is not an exceptional-B value. Since an entire function is a special case of a meromorphic function for which ∞ is exceptional-B, the proof is complete. □

8.5 Theorem. A meromorphic function of finite positive integral order has at most two exceptional-B values.

Proof. Suppose that $f(z)$ is meromorphic and of order ϱ, where ϱ is a positive integer. Suppose further, that the function has three values exceptional-B, which we may take as 0, 1, and ∞. The maximum of their exponents of convergence is $\varrho_1 < \varrho$. Thus

$$f(z) = \frac{P(z)e^{Q(z)}}{H(z)} \qquad \text{and} \qquad f(z) - 1 = \frac{P_1(z)e^{Q_1(z)}}{H(z)}$$

where P, P_1, and H are of order ϱ_1 at most. Thus Q and Q_1 must be polynomials of degree ϱ and

$$P(z)e^{Q(z)} - P_1(z)e^{Q_1(z)} = H(z). \tag{1}$$

If we consider the behavior of the functions as $|z| \to \infty$, we see that the terms of highest degree in Q and Q_1 must be the same, $a_\varrho z^\varrho$, say. Multiplying by $\exp(-a_\varrho z^\varrho)$, we have the left-hand side of (1) of order

$$\max(\varrho_1, \varrho - 1) < \varrho$$

and the right-hand side of order not less than ϱ. Hence we have a contradiction which proves the theorem. □

8.6 FUNCTIONS OF ZERO ORDER

The definition of Borel exceptional values fail for functions of zero order, however, similar methods apply. If $f(z)$ is meromorphic and $f(z) - a$ has only a finite number of zeros, a is a Picard exceptional value (i.e., exceptional-P). If $f(z)$ has only a finite number of poles, ∞ is exceptional-P. We now have the following theorem.

8.6.1 Theorem. A meromorphic function of zero order which is not a rational function has at most one value exceptional-P and an entire function of zero order which is not a polynomial has no finite values exceptional-P.

Proof. Let $f(z)$ be a meromorphic function of zero order, with two values exceptional-P, these values as before may be 0 and ∞. Then $f(z) = P(z)/H(z)$ where P and H are polynomials and thus $f(z)$ is rational. □

Further theorems concerning exceptional values can be studied from the point of view of "Lines of Julia."†

8.7 The main theorems which follow, viz., Picard's "little" and "great" theorems, depend upon a result due to Schottky. This theorem has been considerably expanded in order to illustrate the analysis required. Although it appears that somewhat long-winded theorems and lemmas are needed, all the mathematical apparatus is elementary in nature.

A proof of Picard's theorem using modular functions is considerably more recondite than that presented here.‡

8.7.1 Lemma. Let $\phi(r)$ be a real function of r, for $0 \leq r \leq R_1$ and let $0 \leq \phi(r) \leq M$ for $0 < r \leq R_1$. Further, suppose $\exists$ a constant C such that

$$\phi(r) < C\sqrt{\phi(R)}/(R-r)^2, \qquad 0 < r < R \leq R_1.$$

Then $\exists$ a constant A such that

$$\phi(r) < AC^2/(R_1 - r)^4, \qquad 0 < r < R_1.$$

Proof. The actual form of the result is not particularly important. What is important is that it depends only on r, R_1, and C and not on M.

We have,

$$\phi(r) < C\sqrt{\phi(r_1)}/(r_1 - r)^2$$

† See in particular M. L. Cartwright, "Integral Functions," Chapter VII. Cambridge Univ. Press, London and New York, 1962.

‡ M. E. Picard, "Sur une propriété des fonctions entières," *C. R. Acad. Sci. Paris* **88** (1879) 1024–1027.

for $r < r_1 \leq R_1$. Thus for $0 < r < r_2 < r_1 \leq R_1$,

$$\phi(r_2) < CM^{1/2}/(r_1 - r_2)^2$$

and therefore

$$\phi(r) < \frac{C}{(r_2 - r)^2} \left\{ \frac{C}{(r_1 - r_2)^2} \right\}^{1/2} M^{1/4}.$$

Further,

$$\phi(r) < \frac{C}{(r_3 - r)^2} \left\{ \frac{C}{(r_2 - r_3)^2} \right\}^{1/2} \left\{ \frac{C}{(r_1 - r_2)^2} \right\}^{1/4} M^{1/8}$$

for $0 < r < r_3 < r_2 < r_1 \leq R_1$. Generally,

$$\phi(r) < \frac{C}{(r_n - r)^2} \left\{ \frac{C}{(r_{n-1} - r_n)^2} \right\}^{1/2} \cdots \left\{ \frac{C}{(r_1 - r_2)^2} \right\}^{1/2^{n-1}} M^{1/2^n}$$

for $0 < r_n < r_{n-1} < \cdots < r_1 \leq R_1$. We now put

$$r_n = \tfrac{1}{2}(r + R_1), \qquad r_{n-1} = \tfrac{1}{2}(r_n + R_1), \qquad \ldots$$

$$\phi(r) < 4^{1+1+3/4+\cdots+n/2^{n-1}} \left\{ \frac{C}{(R_1 - r)^2} \right\}^{1+1/2+\cdots+1/2^{n-1}} M^{1/2^n}.$$

As $n \to \infty$, $1 + 1 + \cdots + n/2^{n-1}$ converges, $1 + 1/2 + \cdots + 1/2^{n-1} \to 2$ and $M^{1/2^n} \to 1$, and the result follows. □

8.7.2 Theorem (Schottky). If $f(z)$ is analytic and does not take either of the values 0 or 1 for $|z| \leq R_1$, then for $|z| \leq R < R_1$,

$$|f(z)| < \exp\{KR_1^4/(R_1 - R)^4\}$$

where K depends on $f(0)$ only. For all functions which satisfy the given conditions and are such that $\delta < |f(0)| < 1/\delta$, $|1 - f(0)| > \delta$, K depends on δ only.

Proof. The actual form of the upper bound is not particularly required. This could be considerably improved if necessary. The importance lies in the fact that it depends on $f(0)$ in the manner stated and on R/R_1.

Let $g_1(z) = \log\{f(z)\}$, $g_2(z) = \log\{1 - f(z)\}$ where each log has its principal value at $z = 0$. Thus $g_1(z)$ and $g_2(z)$ are regular for $|z| \leq R_1$ (by hypothesis). Let $M_1(r)$ and $M_2(r)$ be the maxima of $|g_1(z)|$ and

$|g_2(z)|$, respectively, on $|z| = r < R$. Let $M(r) = \max\{M_1(r), M_2(r)\}$. Let

$$B_1(r) = -\min_{|z|=r} \mathscr{R}\{g_1(z)\} = \max_{|z|=r} \log 1/|f(z)|$$

[since for u real, $\max(-u) = -\min(u)$]. We now apply Carathéodory's theorem to $-g_1(z)$ and obtain

$$M_1(\varrho) \leq \frac{2\varrho}{r-\varrho} B_1(r) + \frac{r+\varrho}{r-\varrho} |g_1(0)|, \qquad 0 < \varrho < r. \tag{1}$$

There are now two possibilities: either $B_1(r) \leq 1$, in which case inequality (1) is a result of the required type, or $B_1(r) > 1$ in which case $\exists$ a point z' on $|z| = r$ where $|f(z')|$ is small. However, if $|f(z')|$ is small, $g_2(z)$ is (apart from a term $2n\pi i$) approximately equal to $-f(z')$ [expand $\log(1-f)$]. Thus applying Carathéodory's theorem to $\log g_2$, we have on the left-hand side M_1 (not $\log M_1$ as expected) and on the right-hand side $\log M_2$. If $M_2 < k$, then $\mathscr{R}\{g_2\} < k$ and $|f| < 1 + e^k$, $\forall r$ and we have an equivalent inequality. If $M_2(r) \to \infty$ as $r \to R_1$, it suffices to take $\log M_2 = O(\sqrt{M_2})$ as $M_2 \to \infty$. Suppose $B_1(r) > 1$ and let z' be a point where

$$B_1(r) = \log 1/|f(z')|,$$

then

$$|f(z')| = e^{-B_1(r)} < e^{-1} < \tfrac{1}{2}.$$

Therefore $\exists$ an integer n such that

$$g_2(z') - 2n\pi i = -\sum_{m=1}^{\infty} \{f(z')\}^m/m,$$

hence

$$|g_2(z') - 2n\pi i| \leq \sum_{m=1}^{\infty} |f(z')|^m/m < \sum_{m=1}^{\infty} 2^{-m} = 1.$$

Thus

$$1 > |g_2(z') - 2n\pi i| \geq 2|n|\pi - |g_2(z')|$$

and so

$$2|n|\pi < 1 + |g_2(z')| \leq 1 + M_2(r).$$

Let $h(z) = \log\{g_2(z) - 2n\pi i\}$ where the log has its principal value at $z = 0$. Then $h(z)$ is analytic for $|z| \leq R_1$ since $f(z) \neq 0$, thus $g_2(z) \neq 2n\pi i$.

Carathéodory's theorem then gives

$$\max_{|z|=r} |h(z)| \leq \frac{2r}{R-r} \max_{|z|=r} \log |g_2(z) - 2n\pi i| + \frac{R+r}{R-r} |h(0)|. \tag{2}$$

Now

$$\begin{aligned} |h(z)| = |\log\{g_2(z) - 2n\pi i\}| &= |\log 1/[g_2(z) - 2n\pi i]| \\ &\geq \log |1/[g_2(z) - 2n\pi i]| \end{aligned}$$

hence

$$\max_{|z|=r} |h(z)| \geq \log |1/[g_2(z) - 2n\pi i]|$$

and in particular at z'. Therefore the left-hand side is greater than or equal to

$$\log 1/[|f(z')| + |f(z')|^2 + \cdots] > \log \{1/2 |f(z')|\}$$

and

$$\log |1/[g_2(z') - 2n\pi i]| > B_1(r) - \log 2.$$

On the right-hand side, since

$$|g_2(z) - 2n\pi i| \leq |g_2(z)| + 2|n|\pi \leq M_2(r) + 2|n|\pi,$$

$$\max_{|z|=R} \log |g_2(z) - 2n\pi i| \leq \log\{M_2(R) + 2|n|\pi\} < \log\{2M(R) + 1\}$$

[since $1 + M_2(r) > 2|n|\pi$ and $M(R) \geq M(r)$]. If $n \neq 0$, since

$$g_2(0) = \log\{1 - f(0)\}$$

has its principal value, then $g_2(0) =$ (real part) + i(imaginary part) where $|\text{Imag}| \leq \pi$. Thus

$$|\text{Imag}\ \{g_2(0) - 2n\pi i\}| \geq \pi$$

and

$$|g_2(0) - 2n\pi i| \geq |\text{Imag}\{g_2(0) - 2n\pi i\}| \geq \pi > 1.$$

Since

$$\begin{aligned} h(0) &= \log\{g_2(0) - 2n\pi i\} \\ &= \log |g_2(0) - 2n\pi i| + i \arg\{g_2(0) - 2n\pi i\}, \\ |h(0)| &\leq \big| \log |g_2(0) - 2n\pi i| \big| + |\arg\{g_2(0) - 2n\pi i\}| \\ &\leq \big| \log |g_2(0) - 2n\pi i| \big| + \pi \\ &= \log |g_2(0) - 2n\pi i| + \pi \qquad (\text{since} \quad \log |g_2(0) - 2n\pi i| > 0) \\ &\leq \log\{ |g_2(0)| + 2|n|\pi\} + \pi \\ &\leq \log\{ |g_2(0)| + 1 + M_2(r)\} + \pi. \end{aligned}$$

If $n = 0$,

$$|h(0)| \leq \big|\log|g_2(0)|\big| + \pi.$$

Thus returning to (2),

$$B_1(r) - \log 2 \leq \frac{2R}{R-r}\{\log[2M_2(r)+1] + \log[|g_2(0)| + 1 + M_2(r)] + \pi\}$$

and

$$B_1(r) \leq \frac{2R}{R-r}\{2\log[2M_2(R) + |g_2(0)| + 1] + \big|\log|g_2(0)|\big| + \pi\} + \log 2,$$

since

$$M_2(r) \leq M_2(R) \qquad \text{for} \quad r \leq R,$$
$$2M_2(r) + 1 < 2M_2(R) + |g_2(0)| + 1,$$
$$|g_2(0)| + 1 + M_2(r) \leq |g_2(0)| + 1 + 2M_2(R),$$
$$|h(0)| < \log\{|g_2(0)| + 1 + M_2(R)\} + \pi \qquad \text{for} \quad n = 0$$

or otherwise and $\big|\log|g_2(0)|\big| > 0$. Hence

$$B_1(r) < \frac{4R}{R-r}\{\log[2M_2(R) + |g_2(0)| + 1] + \big|\log|g_2(0)|\big| + \pi\} \tag{3}$$

since $\log 2 < \pi$. The inequality proved now for $B_1(r) > 1$ is clearly true for $B_1(r) \leq 1$ since the right-hand side of (3) exceeds 4π. Combining (1) and (3)

$$\begin{aligned} M_1(\varrho) &< \frac{2\varrho}{r-\varrho} \cdot \frac{4R}{R-r}\{\log[2M_2(R) + |g_2(0)| + 1] \\ &\quad + \big|\log|g_2(0)|\big| + \pi\} + \frac{r+\varrho}{r-\varrho}|g_1(0)|, \\ &< \frac{8rR}{(R-r)(r-\varrho)}\{\log[2M_2(R) + |g_2(0)| + 1] \\ &\quad + \big|\log|g_2(0)|\big| + |g_1(0)| + \pi\} \end{aligned}$$

(since $\varrho < r$). We may interchange g_1 and g_2 in the whole argument because we can interchange f and $1 - f$:

$$1 - f \neq \begin{cases} 1 & \text{if} \ f \neq 0 \\ 0 & \text{if} \ f \neq 1, \end{cases}$$

thus the inequality is true if the suffixes 1 and 2 are interchanged. Hence

$$M_1(\varrho) < \frac{8rR}{(R-r)(r-\varrho)} \{\log[2M_2(R) + |g_2(0)| + 1] + \big|\log|g_2(0)|\big| + |g_1(0)| + \pi\}$$

and

$$M_2(\varrho) < \frac{8rR}{(R-r)(r-\varrho)} \{\log[2M_1(R) + |g_1(0)| + 1] + \big|\log|g_1(0)|\big| + |g_2(0)| + \pi\}.$$

Therefore

$$M(\varrho) < \frac{8rR}{(R-r)(r-\varrho)} \{\log[2M(R) + |g_2(0)| + |g_1(0)| + 1] + \big|\log|g_1(0)|\big| + \big|\log|g_2(0)|\big| + |g_1(0)| + |g_2(0)| + \pi\}$$

or

$$M(\varrho) < \frac{8rR}{(R-r)(r-\varrho)} \{\log M(R) + K_1\},$$

where K_1 depends on $|g_1(0)|$ and $|g_2(0)|$ only. Take $r = \frac{1}{2}(\varrho + R)$ and

$$M(\varrho) < \frac{32R_1^2}{(R-\varrho)^2} \{\log M(R) + K_1\}, \qquad \text{since } R + \varrho < 2R < 2R_1, \text{ etc.}$$

$$< \frac{R_1^2}{(R-\varrho)^2} K_2\sqrt{M(R)}, \qquad \text{since } \log M(R) = O(\sqrt{M(R)}).$$

Thus using the lemma, $M(\varrho) < K_3R_1^4/(R-\varrho)^4$ and we have

$$|f(z)| \leq e^{M(r)} < \exp\{KR_1^4/(R-r)^4\},$$

when K is constant and note $r < \varrho$. Because K depends on $|g_1(0)|$ and $|g_2(0)|$ only, the last part of theorem is true also. □

FURTHER NOTES TO SCHOTTKY'S THEOREM. If $M(R) \approx 0$,

$$\log M(R) \neq O(\sqrt{M(R)}),$$

but if either M_1 or M_2 were bounded, we would immediately obtain $|f|$ bounded.

There is nothing to consider unless $M \to \infty$, when $\log M = O(\sqrt{M})$, and this bound for $\log M$ is taken in order to apply the lemma. We have the following statement: (1) If f is regular (analytic) in $|z| \leq R_1$ and $f \neq 0$, 1 by hypothesis, in $|z| \leq R_1$, then $|f(z)| < B(f(0), |z|)$ in $|z| \leq R_1$. The bound obtained approaches ∞ as $|z| \to R_1$ [not, of course, the actual value of $|f|$, but the bound obtained using this meagre information of the value of $f(0)$]. The bound could be improved, but it would still approach ∞ as $|z| \to R_1$. Since we assume f is analytic in $|z| \leq R_1$, $f(z)$ is bounded there since any continuous function is bounded as such, but there is not a uniform bound for all f satisfying (1) and with the chosen value of $f(0)$. However, there is a uniform bound for all of them in any smaller circle.

The assumption that $f(z)$ does not assume the values 0 or 1 was made only to fix attention, the essence of the theorem remains unchanged if we assume that $f(z)$ does not take any two finite values α and β. It is sufficient then to apply the theorem to the function $G(z) = (f(z) - \alpha)/(\beta - \alpha)$ not taking the values 0 or 1.

8.8 Picard's First Theorem. An entire function which is not constant takes every value, with one possible exception, at least once.

Proof. Suppose that $f(z)$ does not take either of the values a or b ($a \neq b$). Then $g(z) = (f(z) - a)/(b - a)$ does not take either of the values 0 or 1. Thus by Schottky's theorem,

$$|g(z)| < \exp\{KR_1^4/(R_1 - R)^4\}, \qquad |z| \leq R < R_1.$$

Taking $R_1 = 2R$, $|g(z)| < C$. Hence $g(z)$ is constant. □

A generalization of Picard's theorem is as follows.

8.9 Landau's Theorem. If α is any number and β any number not equal to 0, then there is a number $R = R(\alpha, \beta)$ such that every function

$$f(z) = \alpha + \beta z + a_2 z^2 + a_3 z^3 + \cdots,$$

analytic for $|z| \leq R$, takes in this circle one of the values 0 or 1.

Proof. We may suppose $\alpha \neq 0$ or 1 otherwise the result is immediate [since if $\alpha = 0, f(0) = 0$ and if $\alpha = 1, f(0) = 1$]. If $f(z)$ does not take either of the values 0 or 1, then by Schottky's theorem, $f(z) < K(\alpha)$ for $|z| \leq R/2$

because $\alpha = f(0)$. Thus

$$|\beta| = \left|\frac{1}{2\pi i}\int_{|z|=R/2}\frac{f(z)}{z^2}\,dz\right| \leq \frac{K(\alpha)}{R/2} \qquad \text{and hence} \quad R \leq \frac{2K(\alpha)}{|\beta|}.$$

Therefore what is said is, if $f \neq 0, 1$, then the radius of the circle R is less than or equal to a constant. Taking a larger circle, f must take one of the values 0 or 1, i.e., $\exists R(\alpha, \beta)$ such that $\forall f$ of the above form take one of the values 0 or 1 in this circle. The argument uses in effect, contrapositive logic. □

Schottky's theorem saying that, the hypothesis that an analytic function (in $|z| \leq R$) does not assume two values, e.g., 0 or 1, is already a strong limitation on the behavior of the function. Landau's theorem confirms this fact from another direction.

8.9.1 The theorem can be stated in a different way: For every $\alpha, \beta \neq 0$, if the function $f(z) = \alpha + \beta z + a_2 z^2 + \cdots$ is analytic in $|z| \leq R$ and does not assume the values 0 and 1 in it, then $R \leq R(\alpha, \beta)$. Further, since a function has at least one singular point on the circle of convergence we can say the following.

8.9.2 For every α and $\beta \neq 0$, the function $f(z) = \alpha + \beta z + a_2 z^2 + \cdots$ must assume the value 0 or 1 in the closed circle $|z| \leq R(\alpha, \beta)$ or the series has a singular point in this circle.

The assumption $\beta \neq 0$ may be replaced by a more general condition.

8.9.3 Theorem. Let us suppose that $\lambda \neq 0$ and that $f(z) = \alpha + \lambda z^l + a_{l+1} z^{l+1} + \cdots$ is analytic in $|z| \leq R$ and does not assume the values 0 or 1 there. Then $R \leq R(\alpha, \lambda, l)$, i.e., R depends on α, λ, l only.

Proof. As before,

$$|\lambda| = \left|\frac{1}{2\pi i}\int_{|z|=R/2}\frac{f(z)}{z^{l+1}}\,dz\right|$$
$$\leq \frac{1}{2\pi}\,\frac{K(\alpha)}{(R/2)^{l+1}} \cdot 2\pi\,\frac{R}{2} = \frac{K(\alpha)}{(R/2)^l},$$

thus

$$R \leq \frac{2\{K(\alpha)\}^{1/l}}{|\lambda|^{1/l}} = R(\alpha, \lambda, l). \quad \square$$

Papers by Jenkins† deal with the connection between α and β.

Every entire function which is not constant, may be written in the form $f(z) = \alpha + \lambda z^l + a_{l+1}z^{l+1} + \cdots$. Since the radius of convergence is then infinite, the inequality $R \leq R(\alpha, \lambda, l)$ is not satisfied. Thus $f(z)$ must assume at least one of the values 0 or 1. Therefore Landau's theorem includes Picard's "little" theorem in a stronger form, for it gives an estimate of the radius R of a circle (centered at origin) in which $f(z)$ certainly assumes either the value 0 or 1.

8.10 Up to this point, Picard's theorem has been stated in terms of entire functions, i.e., functions with an essential singularity at infinity. A corresponding theorem holds for any function with an isolated essential singularity.

We recall that if $f(z)$ is analytic throughout a neighborhood of a point a, i.e., for $|z - a| < R$, excepting the point a, then a is called an *isolated essential singularity* of the function. Further, if $f(z)$ is analytic in the annulus $R' \leq |z - a| \leq R$, then $f(z)$ may be expanded in a series of positive and negative powers of $(z - a)$, converging at all points of the annulus. Consider the principal part $\sum_{n=1}^{m} b_n(z - a)^{-n}$. If $f(z)$ has a pole at $z = a$, then $|f(z)| \to \infty$ as $z \to a$, for

$$\left| \sum_{n=1}^{m} b_n(z - a)^{-n} \right| = |z - a|^{-m} \left| \sum_{n=1}^{m} b_n(z - a)^{m-n} \right|$$
$$\geq |z - a|^{-m}\left\{ |b_m| - \sum_{n=1}^{m-1} |b_n|\,|z - a|^{m-n}\right\}.$$

As $z \to a$, $\{\cdots\} \to |b_m|$, thus the whole expression approaches ∞, i.e., if $f(z) = O(|z - a|^{-k})$ as $|z - a| \to 0$, the singularity is at most a pole of order k.

8.10.1 Picard's Second Theorem. In the neighborhood of an isolated essential singularity, a single-valued function takes every value with one possible exception, an infinite number of times; or, if $f(z)$ is analytic for $0 < |z - z_0| < \varrho$ and $\exists$ two unequal numbers a, b such that $f(z) \neq a$ and $f(z) \neq b$ for $|z - z_0| < \varrho$, then z_0 is not an essential singularity.

Proof. We may suppose that $z_0 = 0$, $\varrho = 1$, $a = 0$, and $b = 1$. We prove $\exists$ a sequence of circles $|z_n| = r_n$ where $r_n \to 0$ and on which $f(z)$ is bounded.

† J. A. Jenkins, "On explicit bounds in Schottky's theorem," *Canad. J. Math.* **7** (1955) 76–82; "On explicit bounds in Landau's theorem," *ibid.* **8** (1956) 423–425.

This then precludes the existence of a singularity at $z = 0$ except possibly a removable one, for if $f(z) = O(|z - a|^{-k})$ as $|z - a| \to 0$, the singularity is at most a pole of order k and in particular if $f(z) = O(1)$, there is no singularity. Using Weierstrass's theorem which states that in the neighborhood of an isolated essential singularity a function comes arbitrarily close to any given value an infinity of times, there is a sequence of points $z_1, z_2, \ldots$ such that $|z_1| > |z_2| > \cdots > |z_n| \to 0$ and

$$|f(z_n) - 2| < \tfrac{1}{2}.$$

Thus it follows that if all the points are *inside* $|z| = 1$, Schottky's theorem enables us to construct a sequence of circles with these points as centers and in which $f(z)$ is bounded. These circles do not include the origin (since f is not analytic at the origin), however, this is only because Schottky's theorem was proved for the convex curves, viz., circles. The apparent difficulty is removed by making a conformal transformation which replaces the circles by elongated curves which though they exclude the origin, pass around it and overlap on the far side (Fig. 1). Let $z = e^w$ and consider the

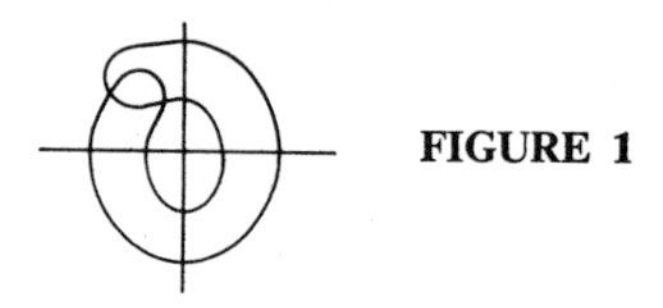

FIGURE 1

half-strip in the w-plane $u < 0$, $-\pi \leq v \leq \pi$: this corresponds to the interior of $|z| = 1$. Let

$$w_n = \log z_n, \qquad -\pi < \mathscr{I}\{w_n\} \leq \pi,$$

Thus

$$\mathscr{R}\{w_n\} = \log |z_n| \to -\infty.$$

Let $f(z) = g(w) = \{f(e^w)\}$. We apply Schottky's theorem to

$$h(w') = g(w_n + w') = f(\exp(w_n + w')).$$

If n is large enough, $z_n \to 0$ and $\exp(w_n + w') \to e^{-\infty} \to 0$. Choose $|w'| \leq 4\pi$, then $h(w')$ is analytic if n is large enough. (Since if $w = w_n + w'$, $|w - w_n| = |w'| \leq 4\pi$ and the circle centered at w_n is sufficiently removed from the origin if the radius is less than or equal to 4π as in Fig. 2. Also $z = e^w = \exp(w_n + w')$, $0 < |z| < 1$). Further, $h(w')$ does not take the values 0 or 1. Thus $|h(w')| < k = k\{h(0)\}$ for $|w'| \leq 2\pi$. Moreover,

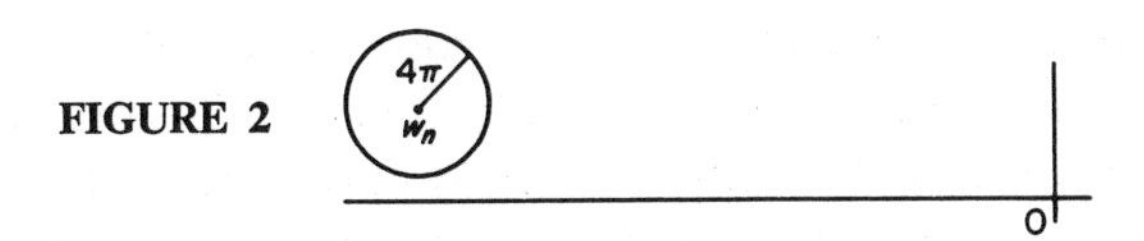

FIGURE 2

$h(0) = g(w_n) = f(z_n)$, satisfying $|f(z_n) - 2| < \frac{1}{2}$, $|z_n| \to 0$. We can replace the right-hand side by an absolute constant. Consequently

$$|g(w_n + w')| = |g(w)| < A \qquad \text{for} \quad |w'| = |w - w_n| \leq 2\pi$$

and in particular for $u = \mathscr{R}\{w_n\}$, $-\pi \leq v \leq \pi$. Thus $|f(z)| < A$, $|z| = |z_n|$. □

8.11 ASYMPTOTIC VALUES

If as $|z| \to \infty$ along some continuous path γ, $f(z) \to a$, then a is an *asymptotic value* of $f(z)$ and γ is a path of *finite determination.* If as $|z| \to \infty$ along γ, $|f(z)| \to \infty$, then infinity is an asymptotic value of $f(z)$ and γ is a path of *infinite determination.* If $|f(z)|$ is bounded on γ but $f(z)$ does not tend to a limit, then γ is a path of *finite indetermination.* If $f(z)$ is meromorphic for $|z| > r$, the definitions also hold.

Zero is an asymptotic value of e^z since $e^z \to 0$ as $|z| \to \infty$ along the negative real axis. Consider $\int_0^z \exp(-t^q)\, dt$, q a positive integer. This function has q asymptotic values

$$\exp(2\pi ik/q) \int_0^\infty \exp(-t^q)\, dt \qquad k = 0, 1, \ldots, q-1 \quad \text{as} \quad z \to \infty$$

along the lines $\arg z = 2\pi k/q$. For transform (rotate) the line of integration $t = re^{2\pi ik/q}$,

$$dt = e^{2\pi ik/q}\, dr$$

and

$$\int_0^z \exp(-t^q)\, dt \to \exp(2\pi ik/q) \int_0^\infty \exp(-r^q \cdot e^{2\pi ik})\, dr \qquad \text{as} \quad |z| \to \infty$$

$$= \exp(2\pi ik/q) \int_0^\infty \exp(-r^q)\, dr.$$

8.11.1 Theorem. Every function with an isolated essential singularity at ∞, and which is not constant, has ∞ as an asymptotic value (or has a path of infinite determination).

Proof. By Laurent's theorem, such a function is of the form $f(z) + g(z)$ where $f(z)$ is an entire function and $g(z)$ tends uniformly to a limit as $|z| \to \infty$. We consider thus entire functions only. The maximum modulus $M(r)$ of a nonconstant entire function has the property that $M(r) \to \infty$ steadily. Consider an indefinitely increasing sequence of numbers $X_1 = M(r_1)$, $X_2 = M(r_2)$, ... Clearly there is a point outside $|z| = r_1$ at which $|f(z)| > X_1$. The set of points where $|f(z)| > X_1$ constitutes the interior of one or more regions bounded by curves on which $|f(z)| = X_1$. These regions must be exterior to $|z| = r_1$. Let one such region be D_1. Then D_1 must extend to ∞ (Fig. 3) otherwise we should have a finite re-

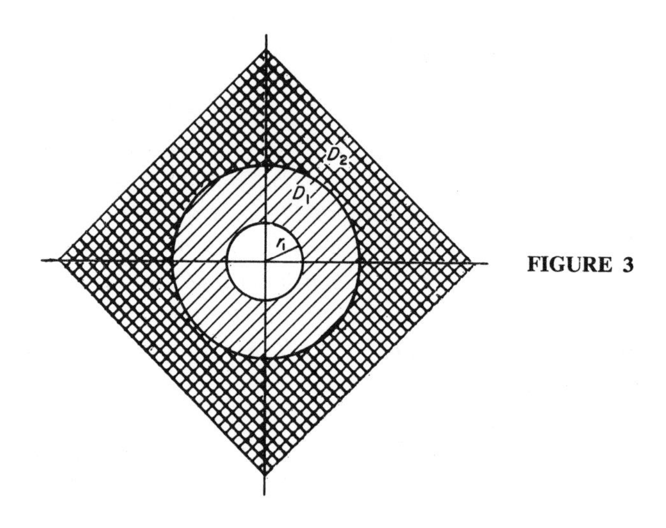

FIGURE 3

gion with $|f(z)| = X_1$ on the boundary and $|f(z)| > X_1$ inside, contrary to the maximum modulus theorem. Also, $f(z)$ is unbounded in D_1 otherwise the Phragmén–Lindelöff principle would show that $|f(z)| \leq X_1$ at all points inside D_1. The Phragmén–Lindelöff argument of Theorem 7.2, applies with P at infinity and $w = r_1/z$. Thus $\exists$ a point of D_1 at which $|f(z)| > X_2$ and consequently a domain D_2 interior to D_1 such that $|f(z)| > X_2$ at all points of D_2. We repeat the argument with X_3, ... Hence $\exists$ a sequence of infinite regions $D_1, D_2, \ldots$ each interior to the preceding one and such that $|f(z)| > X_m$ in D_m and $|f(z)| = X_m$ on the boundary. We take a point on the boundary of D_1 and join it to a point on the boundary of D_2 by a continuous curve lying in D_1, and repeat from D_2 to D_3, and so on. Thus a continuous curve is obtained along which $|f(z)| \to \infty$. □

8.11.2 Theorem. If an entire function does not take the value a (i.e., if a is exceptional-P), then a is an asymptotic value.

Proof. If $f(z)$ is an entire function and $f(z) \neq a$, then $1/(f(z) - a)$ is an entire function and thus by the previous theorem has the asymptotic value ∞; consequently $f(z) \to a$. □

NOTE. The argument in Theorem 7.6 shows that if an entire function has asymptotic values on two curves and is bounded between them, then these asymptotic values must be the same. Asymptotic values not so connected should be considered distinct, whether they are equal or not.

8.12 CONTIGUOUS PATHS

If as $|z| \to \infty$, $f(z) \to a$ on two nonintersecting paths γ_1 and γ_2 starting from the same point z_0, then $\gamma_1 \cup \gamma_2$ determines two infinite domains. If $f(z) \to a$ uniformly in the closure of one of these domains, then γ_1 and γ_2 are defined to be *contiguous paths* of finite determination and the closure of the domain concerned is a *tract* of determination. A similar definition holds for contiguous paths of infinite determination, or finite indetermination. For all other cases, paths of the same kind are defined to be *noncontiguous.*

Clearly, if $f(z) \to a$ as $|z| \to \infty$ along γ_1 and $f(z) \to b$ as $|z| \to \infty$ along γ_2, either $a = b$ or $|f(z)|$ is unbounded between γ_1 and γ_2.

8.12.1 Theorem. Between any two noncontiguous paths of finite determination and between any two noncontiguous paths of finite indetermination, there is a path of infinite determination.

Proof. Suppose $f(z)$ is bounded in one of the domains bounded by $\gamma_1 \cup \gamma_2$ and that $f(z) \to a$ as $|z| \to \infty$ along γ_1 and $f(z) \to b$ along γ_2. By Theorem 7.6 we have that $f(z) \to a$ uniformly in that domain. Thus γ_1 and γ_2 are contiguous paths and we have a contradiction. Similarly, using Theorem 7.2, $|f(z)|$ is unbounded in the case of paths of finite indetermination. The method of Theorem 8.11.1 now gives the result. □

8.12.2 Denjoy conjectured that an entire function of finite order ϱ can have at most 2ϱ different asymptotic values (proved by Ahlfors).[†] We can see, however, that there can be at most 2ϱ straight lines from 0 to ∞ along

[†] L. V. Ahlfors, "Über die asymptotischen Werte der meromorphen Functionen endlicher Ordnung," *Acta. Acad. Abo. Math. Phys.* **6** (1932) 3–8.

which a function of order ϱ has distinct asymptotic values, since by Theorem 7.3 the angle between two such lines must be at least π/ϱ. (N.B. $2\varrho \cdot \pi/\varrho = 2\pi$).

8.12.3 For further results concerning asymptotic values, the reader is directed to a thesis by Al-Katifi.†

† W. Al-Katifi, "On the asymptotic values and paths of certain integral and meromorphic Functions," Imperial College, London (1963) (also in *Proc. London Math. Soc.* (3) **16** (1966) 599–634.

CHAPTER IX

ELEMENTARY NEVANLINNA THEORY

9.1 MEROMORPHIC FUNCTIONS

It will be convenient to use the letters M. F. to mean "Meromorphic Functions" by W. K. Hayman.[†] The theory depends largely on the general Jensen formula, viz., for a function with zeros $a_1, a_2, \ldots a_m$, and poles $b_1, \ldots, b_n$ with moduli not exceeding r and arranged with nondecreasing moduli,

$$\log\left\{\left|\frac{b_1 b_2 \cdots b_n}{a_1 a_2 \cdots a_m} f(0)\right| r^{m-n}\right\} = \frac{1}{2\pi}\int_0^{2\pi} \log|f(re^{i\theta})|\, d\theta .$$

Suppose that in the neighborhood of the origin $f(z) \sim cz^k + \cdots$, where k is any integer. Apply Jensen's formula to $z^{-k}f(z)$.

$$\log\left|\frac{b_1 b_2 \cdots b_n}{a_1 a_2 \cdots a_m}\right| r^{m-n} + \log|c| = \frac{1}{2\pi}\int_0^{2\pi} \log|f(re^{i\theta})|\, d\theta - k\log r.$$

Further,

$$\log \frac{r^m}{|a_1 a_2 \cdots a_m|} = \sum_{\nu=1}^{m-1} \nu \int_{|a_\nu|}^{|a_{\nu+1}|} \frac{dx}{x} + m\int_{|a_m|}^{r} \frac{dx}{x}.$$

[†] W. K. Hayman, "Meromorphic Functions," Oxford Mathematical Monograph. Oxford Univ. Press (Clarendon), London and New York, 1964.

Let $n(r, 0)$ be the number of zeros of $f(z)$ in $|z| \le r$. If $k > 0$,

$$\nu = n(x, 0) - k \text{ for } |a_\nu| \le x < |a_{\nu+1}|$$

(has k zeros at the origin), hence

$$\log \frac{r^m}{|a_1 a_2 \cdots a_m|} = \int_0^r \frac{n(x, 0) - k}{x} dx$$

($= N(r, 1/f)$, M.F. with $f(z)$ such that $k = 0$). If $n(r, \infty)$ is the number of poles of $f(z)$ in $|z| \le r$, then by similar reasoning

$$\log \frac{r^n}{|b_1 b_2 \cdots b_n|} = \int_0^r \frac{n(x, \infty)}{x} dx \qquad (= N(r, f), \quad \text{M.F.}).$$

If $k < 0$, it appears in the second integral instead of the first. Let

$$N(r, a) = \int_0^r \frac{n(x, a) - n(0, a)}{x} dx + n(0, a) \log r$$

[M.F. $N(r, f)$ corresponds to our $N(r, \infty)$, since $n(0, \infty) = 0$]. We obtain,

$$N(r, 0) - N(r, \infty) = \frac{1}{2\pi} \int_0^{2\pi} \log |f(re^{i\theta})| \, d\theta - \log |c| \tag{1}$$

where $n(0, 0) = k$. Now write

$$\log^+ \alpha = \max(\log \alpha, 0), \qquad \alpha > 0.$$

Thus

$$\log \alpha = \log^+ \alpha - \log^+ \frac{1}{\alpha}$$

[since if $\alpha = 1/h$, $h > 1$,

$$\log \alpha < 0, \qquad \log^+ \alpha = \max(\log \alpha, 0) = 0,$$
$$\log \alpha = 0 - \log^+ h = -\log h = \log(1/h)].$$

[Similarly for $\alpha > 1$, $\log^+(1/\alpha) = 0$.] Let

$$m(r, a) \equiv m\left(r, \frac{1}{f - a}\right) = \frac{1}{2\pi} \int_0^{2\pi} \log^+ \left| \frac{1}{f(re^{i\theta}) - a} \right| d\theta$$

and

$$m(r, \infty) \equiv m(r, f) = \frac{1}{2\pi} \int_0^{2\pi} \log^+ | f(re^{i\theta}) | \, d\theta .$$

Then (1) may be written as

$$m(r, 0) + N(r, 0) = m(r, \infty) + N(r, \infty) - \log | c | , \tag{2}$$

since

$$m(r, 0) - m(r, \infty) = \frac{1}{2\pi} \int_0^{2\pi} \log \left| \frac{1}{f(re^{i\theta})} \right| d\theta.$$

We now apply this formula to $f(z) - a$ where a is any number. If $f(z) - a = c_a z^k + \cdots$ in the neighborhood of the origin,

$$m(r, a) + N(r, a) = m(r, f - a) + N(r, \infty) - \log | c_a | , \tag{3}$$

noting that the term $N(r, \infty)$ is unaltered since poles of $f(z) - a$ are the same for each a. {With $a = 0$ this corresponds to the M.F. formula

$$m(r, f) + N(r, f) = m\left(r, \frac{1}{f}\right) + N\left(r, \frac{1}{f}\right) + \log | f(0) | \}.$$

We can of course verify this by the following simple calculation

$$m(r, a) = m\left(r, \frac{1}{\phi - a}\right) = \frac{1}{2\pi} \int_0^{2\pi} \log^+ \left| \frac{1}{\phi(re^{i\theta}) - a} \right| d\theta$$

and $\phi \equiv f - a$.

Applying (2) to ϕ, we have

$$m(r, 0) = m\left(r, \frac{1}{f - a}\right) = \frac{1}{2\pi} \int_0^{2\pi} \log^+ \left| \frac{1}{f(re^{i\theta}) - a} \right| d\theta$$

$$= m\left(r, \frac{1}{\phi}\right) = \frac{1}{2\pi} \int_0^{2\pi} \log^+ \left| \frac{1}{\phi(re^{i\theta})} \right| d\theta$$

and

$$m(r, \infty) = m(r, \phi) = \frac{1}{2\pi} \int_0^{2\pi} \log^+ | \phi(re^{i\theta}) | \, d\theta$$

$$= \frac{1}{2\pi} \int_0^{2\pi} \log^+ | f(re^{i\theta}) - a | \, d\theta = m(r, f - a).$$

Thus

$$m(r, \infty) - m(r, 0)\Big]_\phi = m(r, f - a) - m(r, a)\Big]_f$$

$$= \frac{1}{2\pi} \int_0^{2\pi} \log |f(re^{i\theta}) - a| \, d\theta.$$

$$N(r, 0) = \int_0^r \frac{n(x, 0) - n(0, 0)}{x} dx + n(0, 0) \log r,$$

applied to f. Then $n(x, 0) \to n(x, a)$, since the number of zeros of f in $|z| \leq x$ becomes the number of a-points of f (or zeros of $f - a$) in $|z| \leq x$, and $n(0, 0) \to k$, since $f(z) - a = c_a z^k + \cdots$. Also, $N(r, \infty)$ applied to f becomes $N(r, \infty)$ applied to $f - a$.

Thus

$$N(r, 0) - N(r, \infty)]_\phi = N(r, a) - N(r, \infty)]_f$$

$$= \int_0^r \frac{n(x, a) - k}{x} dx + k \log r - \int_0^r \frac{n(x, \infty)}{x} dx.$$

We observe further, that

$$|f| + |a| \leq 2|fa|, \quad 2|f|, \quad 2|a| \quad \text{or} \quad 2,$$

depending upon whether

$$\text{(i)} \quad |f| \geq 1, \quad |a| \geq 1,$$

$$\text{(ii)} \quad |f| \geq 1, \quad |a| < 1,$$

$$\text{(iii)} \quad |f| < 1, \quad |a| \geq 1,$$

or

$$\text{(iv)} \quad |f| < 1 \quad \text{and} \quad |a| < 1.$$

Verifying, for example, the first result: if $|f| \geq 1$ and $|a| \geq 1$, then $|f||a| \geq |a|$ and $|a||f| \geq |f|$. Thus $|f| + |a| \leq 2|f \cdot a|$. Hence

$$\log\{|f| + |a|\} \leq \log^+ |f| + \log^+ |a| + \log 2. \tag{4}$$

Each case should be verified, e.g., if $|f| \geq 1$, $|a| < 1$, then

$$\log\{|f| + |a|\} \leq \log 2 + \log |f|$$

$$= \log 2 + \log^+ |f|,$$

and $\log^+ | a | = 0$. Since $| f - a | \leq | f | + | a |$,

$$\log^+ | f - a | \leq \log^+ | f | + \log^+ | a | + \log 2 \qquad \text{[from Eq. (4)]}$$

(for if $| f - a | > 1$, $\log^+ | f - a | = \log | f - a |$ and if $| f - a | \leq 1$, $\log^+ | f - a | = 0$ which is certainly less than the right-hand side). Also,

$$\log^+ | f | = \log^+ | (f - a) + a | \leq \log^+\{ | f - a | + | a | \}$$

and

$$\log^+ | f | \leq \log^+ | f - a | + \log^+ | a | + \log 2.$$

Thus

$$\begin{aligned} | m(r, f - a) - m(r, f) | &= \left| \frac{1}{2\pi} \int_0^{2\pi} \{\log^+ | f - a | - \log^+ | f | \} \, d\theta \right| \\ &\leq \frac{1}{2\pi} \int_0^{2\pi} | \{\log^+ | f - a | - \log^+ | f | \} | \, d\theta \\ &\leq \frac{1}{2\pi} \int_0^{2\pi} | \log^+ | a | + \log 2 | \, d\theta \\ &= \log^+ | a | + \log 2. \end{aligned}$$

We now have that since

$$m(r, a) + N(r, a) = m(r, f - a) + N(r, \infty) - \log c_a$$

and

$$m(r, f - a) \leq m(r, \infty) + \log^+ | a | + \log 2,$$

then

$$m(r, a) + N(r, a) = m(r, \infty) + N(r, \infty) + \phi(r, a)$$

where

$$| \phi(r, a) | \leq | \log | c_a | | + \log^+ | a | + \log 2.$$

{This is because

$$m(r, f - a) - \log | c_a | \leq m(r, \infty) + \log^+ | a | + \log 2 - \log | c_a |$$

and

$$m(r, f - a) - \log | c_a | - m(r, \infty) = \phi(r, a)$$

where

$$\phi(r, a) \leq \log^+ | a | + \log 2 - \log | c_a |$$

or

$$| \phi(r, a) | \leq \log^+ | a | + \log 2 + | \log | c_a | | .\}$$

Thus if $f(z)$ is a meromorphic function and not a constant, the value of the sum $m(r, a) + N(r, a)$, for two given values of a, differ by a bounded function of r. Since all the sums are to this extent equivalent, we can represent them all by the one with $a = \infty$. Thus putting

$$T(r) = m(r, \infty) + N(r, \infty)$$

{corresponding to $m(r, f) + N(r, f)$, M.F.} for all a,

$$m(r, a) + N(r, a) = T(r) + \phi(r, a) \tag{5}$$

where $\phi(r, a)$ is (for each a) bounded as $r \to \infty$, and $T(r)$ is called the *characteristic function* of $f(z)$. Then Eq. (5) is called *Nevanlinna's first fundamental theorem.*

9.2 Theorem. $T(r)$ is an increasing convex function of $\log r$.

Proof. We apply Jensen's formula to $f(z) - e^{i\lambda}$ (λ real). For a function $\phi(z)$, with $\phi(0) = c \neq 0$ and $\phi(z)$ analytic in $|z| \leq R$, we have that

$$N(r, 0) - N(r, \infty) = \frac{1}{2\pi} \int_0^{2\pi} \log |\phi(re^{i\theta})| \, d\theta - \log |c|.$$

Substituting in this formula, we note that $N(r, \infty)$ is the same whether we use $f(z)$ or $f(z) - e^{i\lambda}$. $n(x, 0)$ in the formula for $N(r, 0)$ becomes $n(x, e^{i\lambda})$ for $f(z) - e^{i\lambda}$, whereas $n(0, 0)$ applied to $f(z)$ becomes $n(0, e^{i\lambda})$ applied to $f(z) - e^{i\lambda}$. Thus for the function $f(z) - e^{i\lambda}$ we have,

$$N(r, e^{i\lambda}) - N(r, \infty) = \frac{1}{2\pi} \int_0^{2\pi} \log |f(re^{i\theta}) - e^{i\lambda}| \, d\theta - \log |f(0) - e^{i\lambda}| \tag{6}$$

provided $f(0) \neq e^{i\lambda}$. Further, for any value of a,

$$\frac{1}{2\pi} \int_0^{2\pi} \log |e^{i\theta} - a| \, d\theta = \log^+ |a| \tag{7}$$

by applying Jensen's theorem to $f(z) = z - a$, $r = 1$. Note that

$$\log \left\{ \frac{r^n |f(0)|}{r_1 r_2 \cdots r_n} \right\} = \frac{1}{2\pi} \int_0^{2\pi} \log |f(re^{i\theta})| \, d\theta \qquad \text{where} \quad r_n \leq r \leq r_{n+1}.$$

If $a \leq 1$, $\log^+ |a| = 0$ and

$$\log\left\{\frac{r^n |f(0)|}{r_1 r_2 \cdots r_n}\right\} = \log \frac{|-a|}{|a|} = 0.$$

If $a > 1$, $\log^+ |a| = \log a$, and

$$\log\left\{\frac{r^n |f(0)|}{r_1 r_2 \cdots r_n}\right\} = \log |-a|,$$

since if $a > 1$, and $r_1, r_2, \ldots, r_n$ is the product of the moduli of the zeros not exceeding one, the denominator of the left-hand side log is just the number 1, i.e., $\exists$ no zeros of $z - a$ whose moduli are less than or equal to one. In both cases

$$\frac{1}{2\pi}\int_0^{2\pi} \log |e^{i\theta} - a|\, d\theta = \log^+ |a|.$$

We now multiply (6) by $1/2\pi$ and integrate with respect to λ over $[0, 2\pi]$, i.e.,

$$\frac{1}{2\pi}\int_0^{2\pi}\{N(r, e^{i\lambda}) - N(r,\infty)\}d\lambda = \frac{1}{2\pi}\int_0^{2\pi}\left\{\frac{1}{2\pi}\int_0^{2\pi}\log |f(re^{i\theta}) - e^{i\lambda}|\, d\theta\right\}d\lambda$$
$$- \frac{1}{2\pi}\int_0^{2\pi}\log |f(0) - e^{i\lambda}|\, d\lambda.$$

Thus

$$\frac{1}{2\pi}\int_0^{2\pi}\{N(r, e^{i\lambda}) - N(r,\infty)\}\, d\lambda = \frac{1}{2\pi}\int_0^{2\pi}\left\{\frac{1}{2\pi}\int_0^{2\pi}\log |e^{i\lambda} - f(re^{i\theta})|\, d\lambda\right\}d\theta$$
$$- \log^+ |f(0)|$$
$$= \frac{1}{2\pi}\int_0^{2\pi}\log^+ |f(re^{i\theta})|\, d\theta - \log^+ |f(0)|,$$
$$= m(r, \infty) - \log^+ |f(0)|.$$

Hence

$$T(r) = \frac{1}{2\pi}\int_0^{2\pi} N(r, e^{i\lambda})\, d\lambda + \log^+ |f(0)|, \qquad 0 < r < R. \tag{8}$$

Now for any a

$$\frac{d}{d(\log r)}\{N(r, a)\} = r\frac{d}{dr}\{N(r, a)\}$$

and

$$N(r, a) = \int_0^r \frac{n(x, a) - n(0, a)}{x}\, dx + n(0, a) \log r.$$

Therefore

$$\frac{d\{N(r, a)\}}{d(\log r)} = r\left\{\frac{n(r, a) - n(0, a)}{r} + \frac{1}{r}\, n(0, a)\right\} = n(r, a)$$

which is nonnegative and nondecreasing. Thus $N(r, a)$ is an increasing convex function of r (since the second derivative is positive). Thus $T(r)$ has the same property

$$\left\{\frac{d\{T(r)\}}{d(\log r)} = \frac{1}{2\pi} \int_0^{2\pi} \frac{d\{N(r, e^{i\lambda}\}}{d(\log r)}\, d\lambda,\right.$$

and the integrand is nonnegative and increasing}. □

9.3 We also establish a bound for $m(r, a)$ on the circle $|a| = 1$. Using the result (3) with $a = e^{i\theta}$, we have

$$m(r, e^{i\theta}) + N(r, e^{i\theta}) = m(r, f - e^{i\theta}) + N(r, \infty) - \log|f(0) - e^{i\theta}|.$$

Hence

$$\begin{aligned} m(r, e^{i\theta}) + N(r, e^{i\theta}) + \log|f(0) - e^{i\theta}| &= N(r, \infty) + m(r, f - e^{i\theta}) \\ &= T(r) + m(r, f - e^{i\theta}) - m(r, \infty). \end{aligned}$$

Thus

$$T(r) = m(r, e^{i\theta}) + N(r, e^{i\theta}) + \log|f(0) - e^{i\theta}| + G(\theta),$$
$$\text{where} \quad |G(\theta)| \leq \log 2$$

since

$$|m(r, f - e^{i\theta}) - m(r, \infty)| \leq \log^+|e^{i\theta}| + \log 2 = \log 2.$$

We now integrate both sides with respect to θ, from 0 to 2π and using (8),

$$\begin{aligned} \frac{1}{2\pi} \int_0^{2\pi} T(r)\, d\theta = T(r) = \frac{1}{2\pi} \int_0^{2\pi} m(r, e^{i\theta})\, d\theta + T(r) - \log^+|f(0)| \\ + \frac{1}{2\pi} \int_0^{2\pi} \log|f(0) - e^{i\theta}|\, d\theta \\ + \frac{1}{2\pi} \int_0^{2\pi} G(\theta)\, d\theta. \end{aligned}$$

Thus we have

$$0 = \frac{1}{2\pi}\int_0^{2\pi} m(r, e^{i\theta})\, d\theta - \log^+ |f(0)| + \log^+ |f(0)| + \frac{1}{2\pi}\int_0^{2\pi} G(\theta)\, d\theta$$

and

$$\frac{1}{2\pi}\int_0^{2\pi} m(r, e^{i\theta})\, d\theta = -\frac{1}{2\pi}\int_0^{2\pi} G(\theta)\, d\theta \leq \log 2.$$

Consequently $m(r, a)$ is bounded on the average in the circle $|a| = 1$ and a corresponding result holds on any other circle. Thus $N(r, a)$ is a convex increasing function of $\log r$ and so is $T(r)$, however $m(r, a)$ need be neither increasing nor convex in general.

For example, if $f(z) = z/(1 - z^2)$, then $|f(z)| < 1$ for $|z| < \frac{1}{2}$ and $|z| > 2$. Thus $m(r, f) = 0$ for $r \leq \frac{1}{2}$ or $r \geq 2$, however, $f(\pm 1) = \infty$, thus $m(1, f) > 0$.

We note that $N(r, a)$ measures the number of times the function $f(z)$ takes the value a. Since the largest contribution to $m(r, a)$ comes from arcs where $f(z)$ is nearly equal to a, $m(r, a)$ measures in a sense the intensity of the approximation of $f(z)$ to a (or as in M.F., the average smallness in a sense of $f - a$ on $|z| = R$). We could describe $m(r, a) + N(r, a)$ as the total affinity of the function $f(z)$ for the value a. For a given function, certain values may be exceptional, i.e., in the sense that the function does not take these values. The above theorem shows that there can be no exceptional values in the sense that the total affinity of the function for every value is the same, apart from bounded functions of r [since $m(r, a) + N(r, a) = T(r) + \phi$]. We can show that actually, the term $N(r, a)$ predominates.

A few examples will help to consolidate the theory so far.

Example 1. Consider e^z which does not take the value 0 or ∞. These values may be regarded as limiting values as $z \to \pm\infty$. Then

$n(r, 0) = 0,$ since $n(x, 0) = 0$ and $n(0, 0) = 0.$

$N(r, \infty) = 0,$ since e^z has no poles in the finite plane.

$m(r, 0) = \dfrac{r}{\pi},$ since $\log^+ |e^{-z}| = \log^+ e^{-r\cos\theta} = \log e^{-r\cos\theta}$

for $e^{-r\cos\theta} > 1$, i.e.,

$$-r\cos\theta > 0, \qquad r\cos\theta < 0, \qquad \cos\theta < 0, \qquad \pi/2 < \theta < 3\pi/2$$

and

$$\frac{1}{2\pi}\int_{\pi/2}^{3\pi/2} - r\cos\theta\, d\theta = r/\pi,$$

and $m(r, \infty) = r/\pi$ by the same argument. For $a \neq 0$ or ∞,

$$m(r, a) = O(1)$$

and

$$N(r, a) = O(1) + (r/\pi)$$

since

$$m(r, a) + N(r, a) = m(r, \infty) + N(r, \infty) + \phi$$

and ϕ is bounded. Hence

$$T(r) = m(r, \infty) + N(r, \infty) = r/\pi.$$

EXAMPLE 2. Let $f(z)$ be a rational function equal to $P(z)/Q(z)$, with $P(z)$ of degree μ and $Q(z)$ of degree ν, P and Q having no common factor.

a. Suppose $\mu > \nu$. Then $m(r, a) = O(1)$. In either case of $\log^+$, the integral is bounded, thus

$$| Q/(P - aQ) | = O(r^{\nu-\mu}).$$

$$N(r, a) = \mu \log r + O(1),$$

since $dN(r, a)/dr = n(r, a)/r$. And the number of times P/Q assumes the value a is the number of times $P - aQ = 0$, which being a polynomial of degree μ has μ zeros. Thus $dN/dr = \mu/r$ and $N = \mu \log r +$ constant.

$$m(r, \infty) = \frac{1}{2\pi}\int_0^{2\pi} \log^+\left|\frac{P}{Q}\right| d\theta$$

$$= \frac{1}{2\pi}\int_0^{2\pi} \log\{O(r^{\mu-\nu})\}\, d\theta = (\mu - \nu)\log r + O(1)$$

since $\log O(r^{\mu-\nu}) \approx \log cr^{\mu-\nu}$.

$$N(r, \infty) = \nu \log r + O(1),$$

since the number of poles of f in a large enough circle is the degree of Q, viz., ν.

b. For $\mu < \nu$ and $a \neq 0$,

$$m(r, a) = O(1) \qquad \text{since} \quad \frac{1}{|(P/Q) - a|} = O(r^{\nu-\nu}) = O(1).$$

$$N(r, a) = \nu \log r + O(1) \qquad \text{since} \quad n(r, a) = \nu,$$

i.e., the number of zeros of $P - aQ$ is the degree of Q. For $a = 0$,

$$m(r, 0) = (\nu - \mu) \log r + O(1) \qquad \text{and} \quad N(r, 0) = \mu \log r + O(1),$$

since $n(r, 0) =$ degree of P.

c. For $\mu = \nu$, consider the leading coefficient (x^{μ}) in P and Q. Then if $a \neq a_0/b_0$,

$$m(r, a) = O(1) \qquad \text{since} \quad \frac{P}{Q} - a = \frac{a_0}{b_0} + O\left(\frac{1}{r}\right) - a.$$

Then

$$\log \frac{1}{((a_0/b_0) - a) + O(1/r)}$$

is bounded for $r \to \infty$ and $\int_0^{2\pi}$, etc. is bounded.

$$N(r, a) = \mu \log r + O(1),$$

since again $P - aQ$ is of degree $\mu = \nu$.

If $a = a_0/b_0$ $(\mu = \nu)$ we have

$$\frac{1}{(P/Q) - a} = \frac{b_0 Q}{b_0 P - a_0 Q}$$

and

$$\log \left| \frac{b_0 Q}{b_0 P - a_0 Q} \right| = \log c r^{\nu - \alpha},$$

where α is the degree of $b_0 P - a_0 Q$. Therefore

$$m(r, a_0/b_0) = (\mu - \alpha) \log r + O(1).$$

Also,

$$N(r, a_0/b_0) = \alpha \log r + O(1),$$

since the number of times $f = P/Q$ takes the value a_0/b_0 is the number of zeros of $b_0 P - a_0 Q$. In all cases $T(r) = O(\log r)$.

9.4 ORDER OF A MEROMORPHIC FUNCTION

9.4.1 Definition. The meromorphic function $f(z)$ is said to be of order ϱ, if

$$\overline{\lim_{r\to\infty}} \frac{\log T(r)}{\log r} = \varrho$$

so that $T(r) = O(r^{\varrho+\varepsilon})$ for all $\varepsilon > 0$ but not for $\varepsilon < 0$.

We show that this agrees with the definition of order in the case of an entire function.

9.4.2 Theorem. If $f(z)$ is an entire function,

$$T(r) \leq \log^+ M(r) \leq \frac{R+r}{R-r} T(R) \qquad \text{for} \quad 0 < r < R.$$

Proof. Since an entire function has no poles, $N(r, \infty) = 0$. Thus $T(r) = m(r, \infty)$. The left-hand inequality is thus

$$\frac{1}{2\pi} \int_0^{2\pi} \log^+ | f(re^{i\theta}) | \, d\theta \leq \log^+ \max | f(re^{i\theta}) |$$

which is clearly true since $\log^+ \max | f | \geq \log^+ | f |$.

Also, by the Poisson-Jensen formula,

$$\log | f(re^{i\theta}) | = \frac{1}{2\pi} \int_0^{2\pi} \frac{R^2 - r^2}{R^2 - 2Rr\cos(\theta - \phi) + r^2} \log | f(Re^{i\phi}) | \, d\phi$$
$$- \sum_{\mu=1}^{m} \log \left| \frac{R^2 - \bar{a}_\mu re^{i\theta}}{R(re^{i\theta} - a_\mu)} \right|$$

and since $| R^2 - \bar{a}_\mu re^{i\theta} | > | R(re^{i\theta} - a_\mu) |$ because $r < R$ and $R > | a_\mu |$, each term after the integral is negative. Note also, that

$$R^2 - 2R\cos(\theta - \phi) + r^2 \geq (R - r)^2.$$

Thus taking θ so that the left-hand side is a maximum,

$$\log M(r) \leq \frac{R+r}{R-r} \frac{1}{2\pi} \int_0^{2\pi} \log | f(Re^{i\phi}) | \, d\phi \leq \frac{R+r}{R-r} T(R).$$

Thus the right-hand side of the inequality is true. Note, $(R + r)/(R - r) \geq 1$ and $T(R) \geq 0$. $\square$

If we take $R = 2r$, the identity of the two definitions of order of an entire function is clear, since

$$T(r) \leq \log^+ M(r) \leq 3T(2r)$$

and

$$\frac{\log T(r)}{\log r} \leq \frac{\log \log^+ M(r)}{\log r} \leq \frac{\log T(2r)}{\log r} + O\left(\frac{1}{\log r}\right).$$

9.4.3 Now let $r_n(a)$ be the moduli of the zeros of $f(z) - a$, $r_n(\infty)$ the moduli of the poles of $f(z)$, and we have the following results. If $f(z)$ is of order ϱ, then for every a

(i) $m(r, a) = O(r^{\varrho+\varepsilon})$,

(ii) $N(r, a) = O(r^{\varrho+\varepsilon})$,

(iii) $n(r, a) = O(r^{\varrho+\varepsilon})$ and $\sum 1/\{r_n(a)\}^{\varrho+\varepsilon} < \infty$.

We observe that since $m(r, a) = T(r) + O(1) - N(r, a)$ and $N(r, a)$ [and $m(r, a)$] $\nless 0$, that

$$m(r, a) \leq T(r) + O(1) \qquad \text{and} \qquad N(r, a) \leq T(r) + O(1).$$

Thus (i) and (ii) follow {for $T(r) \leq \log^+ M(r) = O(\log \exp r^{\varrho+\varepsilon})$}, and since $n(r, a) = r\, dN(r, a)/dr$, (iii) follows. Also, $f - a$ is a meromorphic function if $f = P/Q$ and the zeros of $f - a$ are the zeros of $P - aQ$ which is an entire function. Thus

$$\sum_{(n)} 1/\{r_n\}^{\varrho+\varepsilon} < \infty$$

as previously proved for entire functions. More precise results can be obtained.†

9.5 FACTORIZATION OF A MEROMORPHIC FUNCTION

Let $f(z)$ be a meromorphic function of order ϱ, with zeros a_n and poles b_n [$f(0) \neq 0$]. Then $\exists$ integers p_1 and p_2 not exceeding ϱ, such that

$$P_1(z) = \prod_{n=1}^{\infty} \left(1 - \frac{z}{a_n}\right) \exp\left\{\frac{z}{a_n} + \cdots + \frac{1}{p_1}\left(\frac{z}{a_n}\right)^{p_1}\right\}$$

† See papers by R. Nevanlinna listed in the Bibliography.

and

$$P_2(z) = \prod_{n=1}^{\infty} \left(1 - \frac{z}{b_n}\right) \exp\left\{\frac{z}{b_n} + \cdots + \frac{1}{p_2}\left(\frac{z}{b_n}\right)^{p_2}\right\}$$

are convergent for all values of z. The function

$$f_1(z) = f(z)P_2(z)$$

is an entire function. Consider

$$f_1(z)\colon\ T(r, f_1) = m(r, \infty, f_1) \leq m(r, \infty, f) + m(r, \infty, P_2)$$

(since $\log^+ fP_2 \leq \log^+ f + \log^+ P_2$) and

$$T(r, f_1) \leq T(r, f) + T(r, P_2) = O(r^{\varrho+\varepsilon}) + O(r^{\varrho+\varepsilon})$$

[P_2 is an entire function and we have proved that $T(r) \leq \log^+ M(r)$]. Thus $f_1(z)$ is of order ϱ at most. Hence $f_1(z) = e^{Q(z)}P_1(z)$ where Q is a polynomial of degree not exceeding ϱ. Thus we have proved that $f(z) = e^{Q(z)}P_1(z)/P_2(z)$, an extension of Hadamard's theorem to meromorphic functions.

9.6 THE AHLFORS–SHIMIZU CHARACTERISTIC

This is now a second formulation of the first fundamental theorem.

9.6.1 Lemma. Let D be a bounded domain, bounded by a finite system of analytic curves γ. Let $f(z)$ be analytic in D and on γ and let $G(R)$ be twice continuously differentiable ($G \in C^2$) on the set of values R assumed by $f(z)$ in D and on γ. Then

$$\int_\gamma \frac{\partial}{\partial n} G(|f(z)|)\, ds = \iint_D g\{|f(re^{i\theta})|\}\, |f'(re^{i\theta})|^2\, r\, dr\, d\theta$$

where

$$g(R) = G''(R) + \frac{1}{R} G'(R),$$

s denotes arc length along γ and $\partial/\partial n$ is differentiation along the normal to γ out of D.

Proof. Consider Green's formula, the two-dimensional analog of the divergence theorem. The divergence theorem states

$$\iiint_V \operatorname{div} \mathbf{F}\, dV = \iint_D \mathbf{F} \cdot \mathbf{n}\, dS$$

If $\mathbf{F} = \mathbf{F}(x, y)$, we choose a cylinder V generated by a line segment of unit length parallel to the z-axis, its lower end describing a contour C. The triple integral now reduces by way of z-integration to $\iint_D \operatorname{div} \mathbf{F}\, dS$, where D is the closed domain in the xy-plane bounded by C. The double integral over the curved surface of V reduces to the line integral around the contour C (after z-integration). Clearly, the integrals over the ends cancel since $\mathbf{n}$ is positive on top and negative on the bottom. Since we now have

$$\iint_D \nabla \cdot \mathbf{F}\, dS = \int_C \mathbf{F} \cdot \mathbf{n}\, ds,$$

choose $\mathbf{F} = \nabla G$ and

$$\nabla \cdot \nabla G = \nabla^2 G, \qquad \nabla \mathbf{G} \cdot \mathbf{n} = \partial G/\partial n$$

and then

$$\iint_D \nabla^2 G r\, dr\, d\theta = \int_\gamma (\partial G/\partial n)\, ds.$$

Taking $G = G(|f|)$, we evaluate $\nabla^2 G(|f|)$ supposing first that $f \neq 0$ in D. We have that $v = \log|f|$ is harmonic in D. Put $|f| = e^v$ and $G(|f|) = G(e^v)$, then

$$\frac{\partial}{\partial x}\{G(e^v)\} = e^v G'(e^v)\frac{\partial v}{\partial x},$$

$$\frac{\partial^2}{\partial x^2}\{G(e^v)\} = e^{2v}G''(e^v)\left(\frac{\partial v}{\partial x}\right)^2 + e^v G'(e^v)\left(\frac{\partial v}{\partial x}\right)^2 + e^v G'(e^v)\frac{\partial^2 v}{\partial x^2}$$

and

$$\frac{\partial^2}{\partial y^2}\{G(e^v)\} = e^{2v}G''(e^v)\left(\frac{\partial v}{\partial y}\right)^2 + e^v G'(e^v)\left(\frac{\partial v}{\partial y}\right)^2 + e^v G'(e^v)\frac{\partial^2 v}{\partial y^2}.$$

Adding and using the fact that v is harmonic,

$$\nabla^2 G(e^v) = \left\{e^{2v}G''(e^v) + e^v G'(e^v)\right\}\left\{\left(\frac{\partial v}{\partial x}\right)^2 + \left(\frac{\partial v}{\partial y}\right)^2\right\}.$$

Writing $e^v = R = |f(z)|$ and noting that

$$\left(\frac{\partial v}{\partial x}\right)^2 + \left(\frac{\partial v}{\partial y}\right)^2 = \left|\frac{d}{dz}\log f\right|^2 = \frac{|f'|^2}{R^2},$$

we have

$$\nabla^2 G(e^v) = \left\{G''(R) + \frac{1}{R} G'(R)\right\} |f'(z)|^2.$$

The result now follows. If f has zeros in D, we exclude them by small circles over which the contribution of $\int (\partial G/\partial n)\, ds$ is negligible since by hypothesis $\partial G(R)/\partial R$ is bounded near $R = 0$. $\square$

We apply the lemma with

$$G(R) = \log \sqrt{1 + R^2}$$

and thus

$$g(R) = \frac{2}{(1 + R^2)^2}.$$

Let $f(z)$ be a meromorphic function in $|z| \leq r$ and suppose f has no poles on $|z| = r$. Exclude poles b_ν of multiplicity k_ν in $|z| < r$, by small circles of radius ϱ (Fig. 1). On such a circle, $f(z) = a_{-k_\nu}/(z - b_\nu)^{k_\nu} + \cdots$. Thus

$$f(z) = \{1/(z - b_\nu)^{k_\nu}\}\phi(z)$$

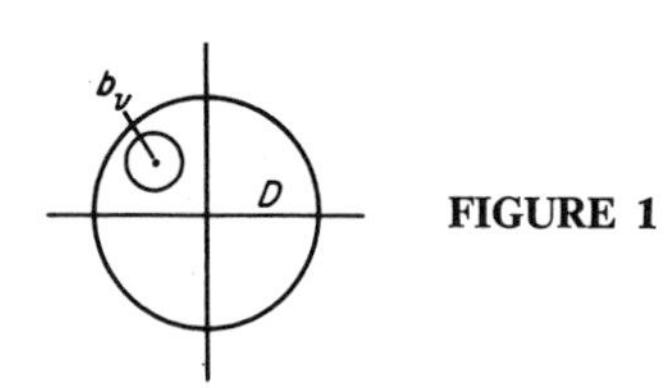

FIGURE 1

where ϕ is analytic in and on $|z - b_\nu| = \varrho$. Hence

$$|f(z)| = (1/\varrho^{k_\nu})\, |\phi| \sim C_\nu/\varrho^{k_\nu}$$

where C_ν is an upper bound for ϕ since ϕ is analytic. Also,

$$\log \sqrt{1 + |f(b_\nu + \varrho e^{i\varphi})|^2} \sim \log |f(b_\nu + \varrho e^{i\varphi})|$$

since $|f| \gg 1$ on the little circle.

$$\log \sqrt{1 + |f(b_\nu + \varrho e^{i\varphi})|^2} = k_\nu \log \frac{1}{\varrho} + O(1)$$

and

$$\frac{\partial}{\partial n} \log \sqrt{1 + |f(b_\nu + \varrho e^{i\varphi})|^2} = \frac{k_\nu}{\varrho} + O(1)$$

since the outward normal to D is directed *into* the little circle. Hence in $\int_\gamma \partial G(R)/\partial n\, ds$, a pole of multiplicity k_ν gives rise to a term $(k_\nu/\varrho)\cdot 2\pi\varrho = 2\pi k_\nu$. Thus

$$\frac{1}{2\pi} r \frac{d}{dr} \int_0^{2\pi} \log\sqrt{1+|f(re^{i\theta})|^2}\, d\theta + \frac{1}{2\pi}\Big(\sum_{(\nu)} 2\pi k_\nu\Big)$$
$$= \frac{1}{2\pi}\int_0^r\int_0^{2\pi} \frac{2\,|f'(\varrho e^{i\theta})|^2}{\{1+|f(\varrho e^{i\theta})|^2\}^2}\, \varrho\, d\varrho\, d\theta,$$

since $\partial/\partial n$ becomes $\partial/\partial r \to d/dr$, ds becomes $r\, d\theta$ and since $\sum_{(\nu)} k_\nu$ becomes $n(r, f)$ the number of poles of f in $|z| < r$ (poles of order p counted p times). Hence

$$\frac{1}{2\pi} r \frac{d}{dr} \int_0^{2\pi} \log\sqrt{1+|f(re^{i\theta})|^2}\, d\theta + n(r, f)$$
$$= \frac{1}{\pi}\int_0^r\int_0^{2\pi} \frac{|f'(\varrho e^{i\theta})|^2}{\{1+|f(\varrho e^{i\theta})|^2\}^2}\, \varrho\, d\varrho\, d\theta.$$

Now call the right-hand side of the previous identity $A(r)$. We divide by r and integrate from 0 to r and obtain

$$\int_0^r \frac{A(t)}{t}\, dt = \frac{1}{2\pi}\int_0^{2\pi} \log\sqrt{1+|f(re^{i\theta})|^2}\, d\theta$$
$$- \log\sqrt{1+|f(0)|^2} + N(r, f) \qquad (9)$$

{using M.F.,

$$N(r, f) = \int_0^r n(t, f)/t\, dt = N(r, \infty)\}.$$

Consider the transformation $W = (1+\bar{a}w)/(w-a)$, $w = f(z)$ and call the resulting transformation $W = F(z)$. Write

$$\frac{1}{k(w, a)} = \sqrt{1+\left|\frac{1+\bar{a}w}{w-a}\right|^2} = \frac{\sqrt{|w-a|^2+|1+\bar{a}w|^2}}{|w-a|}$$
$$= \frac{\sqrt{(1+|a|^2)(1+|w|^2)}}{|w-a|}.$$

$$k(w, a) = \frac{|w-a|}{\sqrt{(1+|a|^2)(1+|w|^2)}} = k(a, w)$$

for w, a both finite and

$$k(a, \infty) = \frac{1}{\sqrt{1 + |a|^2}} = k(\infty, a).$$

Thus $k(w, a) \leq 1$ always. Now

$$\frac{dW}{dz} = -\frac{(1 + |a|^2)}{(w - a)^2} \frac{dw}{dz}$$

and

$$\frac{1}{1 + |W|^2} \left|\frac{dW}{dz}\right| = \frac{(1 + |a|^2)}{|w - a|^2} [k(w, a)]^2 \left|\frac{dw}{dz}\right| = \frac{1}{1 + |w|^2} \left|\frac{dw}{dz}\right|.$$

Thus

$$A(r) = \frac{1}{\pi} \int_0^r \varrho \, d\varrho \int_0^{2\pi} \frac{|f'(\varrho e^{i\theta})|^2}{(1 + |f|^2)^2} \, d\theta = \frac{1}{\pi} \int_0^r \varrho \, d\varrho \int_0^{2\pi} \frac{|F'(\varrho e^{i\theta})|^2}{(1 + |F|^2)^2} \, d\theta.$$

Writing

$$T_0(r) = \int_0^r \frac{A(t)}{t} \, dt$$

and

$$m_0(r, a) = \frac{1}{2\pi} \int_0^{2\pi} \log \frac{1}{k\{f(re^{i\theta}), a\}} \, d\theta$$

we apply (9) to $F(z)$

$$\left[m_0(r, \infty) = \frac{1}{2\pi} \int_0^{2\pi} \sqrt{1 + |f(re^{i\theta})|^2} \, d\theta\right].$$

Clearly,

$$N(r, F(z)) = \int_0^r n(t, F(z)) \frac{dt}{t}$$

$$= N\left(r, \frac{1}{f(z) - a}\right) \equiv N(r, a)$$

since $f(z) - a$ has a zero of order p, at a pole of order p of $F(z)$, for $F(z) = (1 + \bar{a}f(z))/(f(z) - a)$. Thus the first fundamental theorem is obtained in a form due to Ahlfors and Shimizu, viz., the following theorem.

9.6.2 Theorem. If $f(z)$ is meromorphic in $|z| < R$ where $0 < R < \infty$, then for every a finite or infinite, and $0 < r < R$,

$$T_0(r) \equiv \int_0^r A(t)/t\,dt = N(r, a) + m_0(r, a) - m_0(0, a)$$

provided $f(0) \neq a$.

Proof. Since

$$m_0(r, a) = \frac{1}{2\pi} \int_0^{2\pi} \log \sqrt{1 + |F(z)|^2}\, d\theta$$

and

$$m_0(0, a) = \log \sqrt{1 + |F(0)|^2},$$

the proof follows for we have applied the previous result (9) to $F(z)$. □

9.7 We note further, that since

$$\log^+ |f| \leq \log \sqrt{1 + |f|^2} \leq \log^+ |f| + \tfrac{1}{2} \log 2,$$

$$m(r, \infty) \equiv m(r, f) = \frac{1}{2\pi} \int_0^{2\pi} \log^+ |f|\, d\theta \leq m_0(r, \infty) \leq m(r, \infty) + \tfrac{1}{2} \log 2.$$

[We note that if $|f| \leq 1$, $\log^+ |f| = 0$ and $1 + |f|^2 \leq 2$ whereas if $|f| > 1$, $\log^+ |f| = \log |f|$ and $|f|^2 > 1$, $1 + |f|^2 < 2|f|^2$]. Using (9) we have

$$\begin{aligned}
&|T(r) - T_0(r) - \log^+ |f(0)||\\
&\quad = \left| m(r, \infty) + N(r, \infty) - \int_0^r \frac{A(t)}{t}\, dt - \log^+ |f(0)| \right|\\
&\quad = \Big| m(r, \infty) + \log \sqrt{1 + |f(0)|^2}\\
&\qquad - \frac{1}{2\pi} \int_0^{2\pi} \log \sqrt{1 + |f|^2}\, d\theta - \log^+ |f(0)| \Big|\\
&\quad = \left| m(r, \infty) - m_0(r, \infty) + \log \sqrt{1 + |f(0)|^2} - \log^+ |f(0)| \right|\\
&\quad \leq \tfrac{1}{2} \log 2 + \tfrac{1}{2} \log 2 = \log 2.
\end{aligned}$$

Thus $T_0(r)$ and $T(r)$ differ by a bounded term. They may be used interchangeably in most applications. Then $T_0(r)$ is called the *Ahlfors–Shimizu characteristic* and $T(r)$ is called the *Nevanlinna characteristic.*

APPENDIX

We list several definitions, theorems, and observations to complement the introductory chapters and leave it to the reader to consult the standard texts for most of the usual introductory definitions and properties of point sets in the complex plane.

Definition. A nonempty open subset A of the complex plane is *connected* if and only if for every pair of points a, b in A there exists a polygon in A that joins a to b.

Definition. A *domain* is an open connected set.

Definition. A domain D is said to be *simply connected* if every simple closed curve in D contains only points of D in its interior (i.e., D has no "holes" in its interior).

Definition. Let $f(z)$ be a single-valued function defined in a domain D of the complex plane. Let z_0 be any fixed point in D. Then $f(z)$ is said to have a *derivative* at the point z_0 if the number defined by

$$f'(z_0) \equiv \lim_{z \to z_0} \frac{f(z) - f(z_0)}{z - z_0}$$

exists, is finite, and is independent of how $z \to z_0$.

Definition. $f(z)$ is *analytic* (holomorphic) at $z = z_0$ if and only if it is differentiable at every point in some neighborhood of z_0. Thus $f(z)$ is analytic in a domain D if and only if $f(z)$ is analytic at every point of D.

Theorem. A necessary and sufficient condition for a function $f(z) = u(x, y) + iv(x, y)$ to be *analytic* in a domain D is that the four partial derivatives u_x, u_y, v_x, v_y exist, are continuous, and satisfy the Cauchy–Riemann conditions

$$u_x = v_y, \qquad u_y = -v_x$$

at each point of D.

Definition. Given a multivalued function $w = f(z)$ defined in a domain D in the complex plane, let D_0 be a domain contained in D. Then by a *branch* of w in D_0, we mean a single-valued function $w_1 = g(z)$ defined in D_0, such that for any z in D_0, $g(z)$ is one of the values of $f(z)$.

Definition. C is a *continuous arc* if C is the set of all ordered pairs (x, y) such that $x = f(t)$ and $y = g(t)$ where $a \leq t \leq b$ and $f(t)$ and $g(t)$ are continuous.

Definition. C is a *contour* if C is a continuous arc, $f(t)$ and $g(t)$ have sectionally continuous first derivatives and $f(t)$ and $g(t)$ are not zero for the same t.

Definition. C is a *simple* arc (Jordan arc) if (1) C is a continuous arc and (2) if for $t_1 \neq t_2$, then $[f(t_1), g(t_1)] \neq [f(t_2), g(t_2)]$.

Definition. C is a simple *closed* curve if (1) C is a continuous simple arc and (2) for $\alpha \leq t \leq \beta$, $z(\alpha) = z(\beta)$ where $z = z(t) = f(t) + ig(t)$.

Theorem (Jordan curve theorem). Let C be a simple closed curve. Then C divides the plane into two disjoint domains D_1 and D_2. Further, C is the boundary of D_1 as well as D_2. Alternatively, the points not on C form two disjoint domains. Every point of C is an accumulation point of D_1 as well as of D_2. Finally, D_1 and D_2 have no boundary points other than the points of C.

Theorem. Every power series $\sum_{n=0}^{\infty} a_n (z - z_0)^n$ has a radius of convergence R such that when $0 < R < \infty$ the series converges absolutely for $|z - z_0| < R$ and diverges for $|z - z_0| > R$. When $R = 0$, the series converges only for $z = z_0$ and when $R = \infty$, the series converges for all z. The number R is given by

$$\frac{1}{R} = \overline{\lim_{n\to\infty}} \sqrt[n]{|a_n|}.$$

The series may or may not converge on the "circle of convergence," $|z - z_0| = R$.

Definition. A sequence of functions $\{S_n(z)\}$ defined on a set R is said to *converge uniformly* to a function $S(z)$ if, for every $\varepsilon > 0$, there exists a positive integer $N(\varepsilon)$ depending upon ε only, such that for all points z in R

$$|S_n(z) - S(z)| < \varepsilon \qquad \text{for all } n > N(\varepsilon).$$

Theorem. Let the power series $\sum_{n=0}^{\infty} a_n(z - z_0)^n$ have a nonzero radius of convergence R. For any circle Γ, center z_0 and radius $r < R$, the power series $\sum_{n=0}^{\infty} a_n(z - z_0)^n$ converges uniformly within and on Γ.

Theorem (Taylor series). Let $f(z)$ be analytic within a circle C, $|z - z_0| = R$. Then at each point z interior to C,

$$f(z) = \sum_{n=0}^{\infty} \frac{f^{(n)}(z_0)}{n!} (z - z_0)^n.$$

The series converges and has $f(z)$ as its sum function.

Theorem (Laurent series). Let S be the region bounded by the concentric circles C_1 and C_2 with center z_0 and radii r_1 and r_2, respectively, $r_1 < r_2$. Let $f(z)$ be analytic within S and on C_1 and C_2. Then at each point z in the interior of S, $f(z)$ can be represented by a convergent series of positive and negative powers of $(z - z_0)$,

$$f(z) = \sum_{n=0}^{\infty} a_n(z - z_0)^n + \sum_{n=1}^{\infty} b_n(z - z_0)^{-n},$$

where

$$a_n = \frac{1}{2\pi i} \int_{C_2} \frac{f(\xi)}{(\xi - z_0)^{n+1}} \, d\xi, \qquad n = 0, 1, 2, \ldots$$

and

$$b_n = \frac{1}{2\pi i} \int_{C_1} \frac{f(\xi)}{(\xi - z_0)^{-n+1}} \, d\xi, \qquad n = 1, 2, \ldots$$

the integral along C_1 and C_2 being taken in the positive direction.

SUGGESTIONS FOR FURTHER READING

Boas, R. P., Jr. "Entire Functions." Academic Press, New York, 1954.

Branges, L. de, "Hilbert Spaces of Entire Functions." Prentice-Hall, Englewood Cliffs, New Jersey, 1968.

Cartwright, M. L. "Integral Functions." Cambridge Univ. Press, London and New York, 1962.

Gross, F., ed., *Proc. NRL Conf. Classical Function Theory*. Math. Res. Center, Naval Res. Lab., Washington, D. C., 1970.

Hayman, W. K. "Meromorphic Functions." Oxford Univ. Press (Clarendon), London and New York, 1964.

Markushevich, A. I. "Entire Functions." American Elsevier, New York, 1966.

Tsuji, M. "Potential Theory in Modern Function Theory." Maruzen, Tokyo, 1959.

Valiron, G. "Theory of Integral Functions." Chelsea, New York, 1949.

Levin, B. Ja. Distribution of Zeros of Entire Functions, A.M.S. Translations of Math. Mono. No. 5, 1964.

BIBLIOGRAPHY

The following moderately expanded bibliography covers not only the subjects touched on in this monograph, but extensions of theorems proved herein. The subject of entire functions is far too wide to be able to produce an exhaustive set of references, however an attempt has been made to give a reasonably comprehensive cross section of papers and books relating to the subject.

AGMON, S.

(1) "Functions of exponential type in an angle and singularities of Taylor series." *Trans. Amer. Math. Soc.* **70** (1951) 492–508.

AHIEZER, N. I. (AKHYESER, AKHIEZER, ACHYESER, etc.)

(1) "Sur les fonctions entières d'ordre entier." *Rend. Circ. Mat. Palermo* (1) **51** (1927) 390–393.

(2) "On some properties of integral transcendental functions of exponential type." *Izv. Akad. Nauk SSSR Ser. Math.* **10** (1946) 411–428 (Russian; English summary).

(3) "Lectures on the Theory of Approximation." Ogiz, Moscow and Leningrad, 1947 (Russian); (translation: Ungar, New York, 1956).

(4) "On the theory of entire functions of finite degree." *Dokl. Akad. Nauk SSSR* (*N.S.*) **63** (1948) 475–478 (Russian).

(5) "On the interpolation of entire transcendental functions of finite degree." *Dokl. Akad. Nauk SSSR* (*N.S.*) **65** (1949) 781–784 (Russian).

(6) "The work of Academician S. N. Bernstein on the constructive theory of functions." *Usp. Matem. Nauk* (*N.S.*) **6**, No. 1 (41) (1951) 3–67 (Russian).

(7) "On entire transcendental functions of finite degree having a majorant on a sequence of real points." *Izv. Akad. Nauk SSSR. Ser. Mat.* **16** (1952) 353–364 (Russian).

AHLFORS, L.
(1) "Über die Kreise, die von einer Riemannschen Fläche schlicht überdeckt werden." *Comm. Math. Helvetici* **5** (1933) 28–38.
(2) "Sur les domains dans lesquels une fonction méromorphe prend des valeurs appartenant à une region donnée." *Acta Sci. Fennicae Nova Ser. A II* **2** (1933) 1–17.
(3) "Zur Theorie der Überlagerungsflächen." *Acta Math.* **65** (1935) 157–194.
(4) "Beiträge zur Theorie der meromorphen Funktionen." *C.R. 7e Congr. Math. Scand. Oslo* (1929) 84–88.
(5) "Ein Satz von Henri Cartan und seine Anwendung auf die Theorie der meromorphen Funktionen." *Soc. Sci. Fenn. Comment. Phys. Math.* (16) **5** (1931).
(6) "An extension of Schwarz's lemma." *Trans. Amer. Math. Soc.* **43** (1938) 359–364.
(7) "Untersuchungen zur Theorie der konformen Abbildung und der ganzen Funktionen." *Acta Soc. Sci. Fenn. Ser. A* **1** (9) (1930).
(8) "On Phragmén-Lindelöf's Principle." *Trans. Amer. Math. Soc.* **41** (1937) 1–8.
(9) "Über die asymptotischen Werte der meromorphen Funktionen endlicher Ordnung." *Acta Acad. Abo. Math. Phys.* **6** (9) (1932) 3–8.
(10) "Über eine Methode in der Theorie der meromorphen Funktionen." *Soc. Sci. Fenn. Comment. Phys.-Math.* **8** (1935).

AHLFORS, L. V., and GRUNSKY, H.
(1) "Über die Blochsche Konstante." *Math. Z.* **42** (1937) 671–673.

AHLFORS, L., and HEINS, M.
(1) "Questions of regularity connected with the Phragmén-Lindelöf principle." *Ann. Math.* (2) **50** (1949) 341–346.

AHLFORS, L. V., and SARIO, L.
(1) "Riemann Surfaces." Princeton Univ. Press, Princeton, New Jersey (1960).

ALANDER, M.
(1) "Sur les fonctions entières non-réelles." *Ark. Mat., Astron. Fysik* **18** (1924) 1–9.

AL-KATIFI, W.
(1) "On the asymptotic values and paths of certain integral and meromorphic functions." *Proc. London Math. Soc.* (3) **16** (1966) 599–634.

AMIRÀ, B.
(1) "Sur un théorème de M. Wiman dans la théorie des fonctions entières." *Math. Z.* **22** (1925) 206–221.

ARIMA, K.
(1) "On maximum modulus of integral functions." *J. Math. Soc. Japan* **4** (1952), 62–66.

BAKER, I. N.
(1) "Sets of non-normality in iteration theory." *J. London Math. Soc.* **40** (1965) 499–502.
(2) "The distribution of fixpoints of entire functions." *Proc. London Math. Soc.* (3) **16** (1966) 493–506.
(3) "Zusammensetzungen ganzer Funktionen." *Math. Z.* **69** (1958) 121–163.
(4) "The existence of fixpoints of entire functions." *Math. Z.* **73** (1960) 280–284.
(5) "Repulsive Fixpoints of Entire Functions." *Math. Z.* **104** (1968) 252–256.
(6) "Permutable entire functions." *Math. Z.* **79** (1962) 243–249.

(7) "On some results of A. Rényi and C. Rényi concerning periodic entire functions." *Acta Sci. Math.* **27** (1966) 197–200.
(8) "Entire functions with linearly distributed values." *Math. Z.* **86** (1964) 263–267.
(9) "Fixpoints of polynomials and rational functions." *J. London Math. Soc.* **39** (1964) 615–622.
(10) "Fixpoints and iterates of entire functions." *Math. Z.* **71** (1959) 146–153.
(11) "Some entire functions with fixpoints of every order." *J. Austr. Math. Soc.* (to appear).
(12) "Multiply connected domains of normality in iteration theory." *Math. Z.* **81** (1963) 206–214.

BAKER, I. N., and GROSS, F.
(1) "On Factorizing Entire Functions." *Proc. London Math. Soc.* (3) **18** (1968) 69–76.

BELLMAN, R.
(1) "A generalization of a Zygmund-Bernstein theorem." *Duke Math. J.* **10** (1943) 649–651.

BERNSTEIN, S.
(1) "Sur une propriété des fonctions entières." *C. R. Acad. Sci. Paris* **176** (1923) 1603–1605.
(2) "Leçons sur les propriétés extrémales et la meilleure approximation des fonctions analytiques d'une variable réelle." Gauthier-Villars, Paris, 1926.
(3) "Sur la meilleure approximation de $|x|^p$ par des polynômes de degrés très élevés." *Izv. Akad. Nauk SSSR Ser. Mat.* (1938), 169–190 (Russian and French).
(4) "Sur la borne supérieure du module de la dérivée d'une fonction de degré fini." *Compt. Rend. Dokl. Acad. Sci. URSS (N.S.)* **51** (1946) 567–568.
(5) "Sur la meilleure approximation sur tout l'axe réel des fonctions continues par des fonctions entières de degré fini, I, II, III, IV, V." *Compt. Rend. Dokl. Acad. Sci. URSS (N.S.)* **51** (1946) 331–334, 487–490; **52** (1946) 563–566; **54** (1946) 103–108, 476–478.
(6) "Démonstration nouvelle et généralisation de quelques formules de la meilleure approximation." *Compt. Rend. Dokl. Acad. Sci. URSS (N.S.)* **54** (1946) 663–664.
(7) "On limiting relations among constants of the theory of best approximation." *Dokl. Akad. Nauk SSSR (N.S.)* **57** (1947) 3–5 (Russian).
(8) "On properties of homogeneous functional classes." *Dokl. Akad. Nauk SSSR (N.S.)* **57** (1947) 111–114 (Russian).
(9) "Limit laws of the theory of best approximation." *Dokl. Akad. Nauk SSSR (N.S.)* **58** (1947) 525–528 (Russian).
(10) "A second note on homogeneous functional classes." *Dokl. Akad. Nauk SSSR (N.S.)* **59** (1948) 1379–1384 (Russian).
(11) "A generalization of an inequality of S. B. Stechkin to entire functions of finite degree." *Dokl. Akad. Nauk SSSR (N.S.)* **60** (1948) 1487–1490 (Russian).
(12) "The extension of properties of trigonometric polynomials to entire functions of finite degree." *Izv. Akad. Nauk SSSR. Ser. Mat.* **12** (1948) 421–444 (Russian).
(13) "Remarks on my paper "The extension of properties of trigonometric polynomials to entire functions of finite degree." *Izv. Akad. Nauk SSSR. Ser. Mat.* **12** (1948) 571–573 (Russian).
(14) "On additive majorants of finite growth." *Dokl. Akad. Nauk SSSR (N.S.)* **66** (1949) 545–548 (Russian).

(15) "On some new results in the theory of approximation of functions of a real variable." *Acta Sci. Math.* (*Szeged*) **12** Leopoldo Fejér et Frederico Riesz LXX annos natis dedicatus, Part A (1950) 161–169 (Russian).
(16) "On weight functions." *Dokl. Akad. Nauk SSSR* (*N.S.*) **77** (1951) 549–552 (Russian).

BERNSTEIN, V.
(1) "Leçons sur les progrès recents de la théorie des séries de Dirichlet." Gauthier-Villars, Paris, 1933.
(2) "Sopra una proposizione relativa alla crescenza delle funzioni olomorfe." *Ann. Scuola Norm. Super. Pisa* (2) **2** (1933) 381–399.

BESICOVITCH, A. S.
(1) "On integral functions of order < 1." *Math. Ann.* **97** (1927) 677–695.

BEURLING, A.
(1) "Some theorems on boundedness of analytic functions." *Duke Math. J.* **16** (1949) 355–359.

BIEBERBACH, L.
(1) "Über eine Vertiefung des Picardschen Satzes bei ganzen Funktionen endlicher Ordnung." *Math. Z.* **3** (1919) 175–190.
(2) "Zwei Sätze über das Verhalten analytischer Funktionen in der Umgebung wesentlich singulärer Stellen." *Math. Z.* **2** (1918) 158–170.
(3) "Über einen Satz Pólyascher Art." *Arch. Math.* **4** (1953) 23–27.
(4) "Lehrbuch der Funktionentheorie," vol. 1, 3rd ed., and vol. 2, 2nd ed. Tuebner, Leipzig, 1930, 1931.
(5) "Auszug aus einem Briefe des Herrn Bieberbach an den Herausgeber." *Acta Math.* t. 42.

BIERNACKI, M.
(1) "Sur le déplacement des zéros des fonctions entières par leur dérivation." *C. R. Acad. Sci. Paris* **175** (1922) 18–20.
(2) "Sur les directions de Borel des fonctions méromorphes." *Acta Math.* **56** (1930), 197–204.

BLASCHKE, W.
(1) "Eine Erweiterung des Satzes von Vitali über Folgen analytischer Funktionen." *S.-B. Leipziger Akad. Wiss.* **67** (1915) 194–200.
(2) "Kreis und Kugel," 2nd ed. De Gruyter Berlin, 1956.

BLOCH, A.
(1) "Les théorèmes de M. Valiron sur les fonctions entières et la théorie de l'uniformisation." *Ann. Fac. Sci. Univ. Toulouse* (3) **17** (1926) 1–22.

BLUMENTHAL, O.
(1) "Über ganze transzendente Funktionen." *D. Math. Ver.* t. 16.
(2) "Sur le mode de croissance des fonctions entières." *Bull. Soc. Math.* (1907).
(3) "Principes de la théorie des fonctions entières d'ordre infini." Gauthier-Villars, Paris, 1910.

BOAS, R. P., JR.
(1) "Entire Functions." Academic Press, New York, 1954.
(2) "Some theorems on Fourier transforms and conjugate trigonometric integrals." *Trans. Amer. Math. Soc.* **40** (1936) 287–308.

(3) "The derivative of a trigonometric integral." *J. London Math. Soc.* **12** (1937) 164–165.

(4) "Asymptotic relations for derivatives." *Duke Math. J.* **3** (1937) 637–646.

(5) "Representations for entire functions of exponential type." *Ann. Math.* (2) **39** (1938) 269–286; correction, *ibid.* **40** (1939) 948.

(6) "Remarks on a theorem of B. Lewitan." *Rec. Math.* (*Mat. Sb.*) (*N.S.*) **5** (47) (1939) 185–187.

(7) "Entire functions bounded on a line." *Duke Math. J.* **6** (1940) 148–169; correction, *ibid.* **13** (1946) 483–484.

(8) "Some uniqueness theorems for entire functions." *Amer. J. Math.* **62** (1940) 319–324.

(9) "Univalent derivatives of entire functions." *Duke Math. J.* **6** (1940) 719–721.

(10) "Expansions of analytic functions." *Trans. Amer. Math. Soc.* **48** (1940) 467–487.

(11) "A note on functions of exponential type." *Bull. Amer. Math. Soc.* **47** (1941) 750–754.

(12) "Entire functions of exponential type." *Bull. Amer. Math. Soc.* **48** (1942) 839–849.

(13) "Representation of functions by Lidstone series." *Duke Math. J.* **10** (1943) 239–245.

(14) "Functions of exponential type, I." *Duke Math. J.* **11** (1944) 9–15.

(15) "Functions of exponential type, III." *Duke Math. J.* **11** (1944) 507–511.

(16) "Fundamental sets of entire functions." *Ann.* (2) **47** (1946) 21–32. correction, *ibid.* **48** (1947) 1095.

(17) "The growth of analytic functions." *Duke Math. J.* **13** (1946) 471–481.

(18) "Quelques généralisations d'un théorème de S. Bernstein sur la dérivée d'un polynome trigonométrique." *C. R. Acad. Sci. Paris* **227** (1948) 618–619.

(19) "Exponential transforms and Appell polynomials." *Proc. Nat. Acad. Sci. U.S.A.* **34** (1948) 481–483.

(20) "Sur les séries et intégrales de Fourier a coefficients positifs." *C. R. Acad. Sci. Paris* **228** (1949) 1837–1838.

(21) "Polynomial expansions of analytic functions." *J. Indian Math. Soc.* **14** (1950) 1–14.

(22) "Differential equations of infinite order." *J. Indian Math. Soc.* **14** (1950) 15–19.

(23) "Partial sums of Fourier series." *Proc. Nat. Acad. Sci. U.S.A.* **37** (1951) 414–417.

(24) "Growth of analytic functions along a line." *Proc. Nat. Acad. Sci. U.S.A.* **38** (1952) 503–504.

(25) "Integrability along a line for a class of entire functions." *Trans. Amer. Math. Soc.* **73** (1952) 191–197.

(26) "Inequalities between series and integrals involving entire functions." *J. Indian Math. Soc.* **16** (1952) 127-135.

(27) "Oscillation of partial sums of Fourier series." *J. Analyse Math.* **2** (1952) 110–126.

(28) "Integral functions with negative zeros." *Can. J. Math.* **5** (1953) 179–184.

(29) "Two theorems on integral functions." *J. London Math. Soc.* **28** (1953) 194–196.

(30) "A Tauberian theorem for integral functions." *Proc. Cambridge Philos. Soc.* **49** (1953) 728–730.

(31) "Asymptotic properties of functions of exponential type." *Duke Math. J.* **20** (1953) 433–448.

(32) "Growth of analytic functions along a line." *J. Analyse Math.* **4** (1954–1956) 1–28.

BOAS, R. P., JR., BUCK, R. C., and ERDÖS, P.
(1) "The set on which an entire function is small." *Amer. J. Math.* **70** (1948) 400–402.

BOAS, R. P., JR., and POLLARD, H.
(1) "Complete sets of Bessel and Legendre functions." *Ann. Math.* (2) **48** (1947) 366–383.

BOAS, R. P., JR., and SCHAEFFER, A. C.
(1) "A theorem of Cartwright." *Duke Math. J.* **9** (1942) 879–883.

BOHR, H.
(1) "Über einen Satz von Edmund Landau." *Scripta Univ. Hierosolymitanarum* **1** (2) (1923) 1–5.

BONNESEN, T., and FENCHEL, W.
(1) "Theorie der konvexen Körper. Ergebnisse der Mathematik und ihrer Grenzgebiete," Vol. 3, No. 1. Springer-Verlag, Berlin, 1934.

BOREL, E.
(1) "Sur les zéros des fonctions entières." *Acta Math.* **20** (1897) 357–396.
(2) "Leçons sur les fonctions entières." Gauthier-Villars, Paris, 1900.
(3) "Leçons sur les fonctions entières. Deuxieme édition revue et augmentée d'une note de G. Valiron." Gauthier-Villars, Paris, 1921.
(4) "Démonstration élémentaire d'un théorème de M. Picard sur les fonctions entières." *C. R. Acad Sci. Paris.* **122** (1896) 1045–1048.
(5) "Sur les fonctions entières de genre infini." *C. R. Acad. Sci. Paris.* **134** (1902) 1343–1344.
(6) "Leçons sur les séries a termes positifs." Gauthier-Villars, Paris, 1902.
(7) "Leçons sur les fonctions méromorphes." Gauthier-Villars, Paris, 1903.
(8) "Leçons sur la théorie de la croissance." Gauthier-Villars, Paris, 1910.
(9) "Problèmes et méthodes de théorie des fonctions." Gauthier-Villars, Paris, 1922.
(10) "Sur quelques fonctions entières." *Rend. Circ. Mat. Palermo*, t. 23.

BOUTROUX, P.
(1) "Sur quelques propriétés des fonctions entières." Thèse. *Acta Math.* 1903.
(2) "Sur l'indétermination d'une fonction holomorphe dans le voisinage d'une singularité essentielle." *Ann. Sci. École Norm. Sup.* 1908.

BOWEN, N. A.
(1) "A function-theory proof of Tauberian theorems on integral functions." *Quart. J. Math. Oxford Ser.* (1) **19** (1948) 90–100.

BOWEN, N. A., and MACINTYRE, A. J.
(1) "An oscillation theorem of Tauberian type." *Quart. J. Math. Oxford Ser.* (2) **1** (1950) 243–247.
(2) "Some theorems on integral functions with negative zeros." *Trans. Amer. Math. Soc.* **70** (1951) 114–126.

BRANGES, L. DE
(1) "Hilbert Spaces of Entire Functions." Prentice-Hall, Englewood Cliffs, New Jersey, 1968.
(2) "Some mean squares of entire functions." *Proc. Amer. Math. Soc.* **10** (1959) 833–839.
(3) "Some Hilbert spaces of entire functions." *Proc. Amer. Math. Soc.* **10** (1959) 840–846.
(4) "Some Hilbert spaces of entire functions." *Trans. Amer. Math. Soc.* **96** (1960) 259–295.

(5) "Some Hilbert spaces of entire functions II." *Trans. Amer. Math. Soc.* **99** (1961) 118–152.
(6) "Some Hilbert spaces of entire functions III." *Trans. Amer. Math. Soc.* **100** (1961) 73–115.
(7) "Some Hilbert spaces of entire functions IV." *Trans. Amer. Math. Soc.* **105** (1962) 43–83.
(8) "Homogeneous and periodic spaces of entire functions." *Duke Math. J.* **29** (1962) 203–224.
(9) "Symmetry in spaces of entire functions." *Duke Math. J.* **29** (1962) 383–392.
(10) "Perturbations of self-adjoint transformations." *Amer. J. Math.* **84** (1962) 543–560.
(11) "Some applications of spaces of entire functions." *Canad. J. Math.* **15** (1963) 563–583.
(12) "A comparison theorem for spaces of entire functions." *Proc. Amer. Math. Soc.* **14** (1963) 445–468.
(13) "New and old problems for entire functions." *Bull. Amer. Math. Soc.* **70** (1964) 214–223.
(14) "Some examples of spaces of entire functions." *J. Reine Angew. Math.* **222** (1966) 20–54.
(15) "The expansion theorem for Hilbert spaces of entire functions." *Proc. Symp. Pure Math.* Vol. XI, "Entire Functions and Related Parts of Analysis." American Mathematical Society, Providence, Rhode Island, 1968, pp. 79–148.
(16) "Kummer spaces of entire functions." Unpublished manuscript, 1967.

BRINKMIER, H.
(1) "Über das Mass der Bestimmutheit des Wachstums einer ganzen transzendenten Function durch die absoluten Betrage der Koeffizienten ihrer Potenzreihe." *Math. Ann.* **96** (1927) 108–718.

BRUNK, H. D.
(1) "On the growth of functions having poles or zeros on the positive real axis." *Pacific J. Math.* **4** (1954) 1–19.

BUCK, R. C.
(1) "An extension of Carlson's theorem." *Duke Math. J.* **13** (1946) 345–349.
(2) "A class of entire functions." *Duke Math. J.* **13** (1946) 541–559.
(3) "Interpolation and uniqueness of entire functions." *Proc. Nat. Acad. Sci. U.S.A.* **33** (1947) 288–292.
(4) "Interpolation series." *Trans. Amer. Math. Soc.* **64** (1948) 283–298.
(5) "Integral valued entire functions." *Duke Math. J.* **15** (1948) 879–891.
(6) "On the distribution of the zeros of an entire function." *J. Indian Math. Soc.* **16** (1952) 147–149.
(7) "On admissibility of sequences and a theorem of Pólya." *Comment. Math. Helv.* **27** (1953) 75–80.

CAMERON, R. H., and MARTIN, W. T.
(1) "Infinite linear differential equations with arbitrary real spans and first degree coefficients." *Trans. Amer. Math. Soc.* **54** (1943) 1–22.

CARLEMAN, T.
(1) "Über die Approximation analytischer Funktionen durch lineare Aggregate von vorgegebenen Potenzen." *Ark. Mat. Astr. Fys.* **17** (9) (1922) 1–30.

(2) "Sur un théorème de Weierstrass." *Ark. Mat. Astr. Fys.* **20B** (4) (1927).
(3) "Les Fonctions quasi-analytiques." Borel Monograph, Paris 1926.
(4) "Sur les fonctions inverses des fonctions entières d'ordre fini." *Arkiv. Math.* (10) t. 15.

CARLESON, L.

(1) "On infinite differential equations with constant coefficients, I." *Math. Scand.* **1** (1953) 31–38.

CARMICHAEL, R. D.

(1) "Linear differential equations of infinite order." *Bull. Amer. Math. Soc.* **42** (1936) 193–218.

CARTAN, H.

(1) "Sur la fonction de croissance attachée à une fonction méromorphe de deux variables, et ses applications aux fonctions méromorphes d'une variable." *C. R. Acad. Sci. Paris* **189** (1929) 521–523.
(2) "Sur les systèmes de fonctions holomorphes a variétés linéaires lacunaires et leurs applications." *Ann. Sci. École Norm. Sup.* (3) **45** (1928) 255–346.

CARTWRIGHT, M. L.

(1) "Some inequalities in the theory of functions." *Math. Ann.* **111** (1935) 98–118.
(2) "The radial limits of functions meromorphic in a circular disk." *Math. Z.* **76** (1961) 404–410.
(3) "Integral functions." Cambridge Univ. Press, London and New York, 1962 (Camb. Tracts in Math. and Math. Phys., No. 44).
(4) "The zeros of certain integral functions." *Quart. J. Math. Oxford Ser.* (1) **1** (1930) 38–59.
(5) "The zeros of certain integral functions, II." *Quart. J. Math. Oxford Ser.* (1) **2** (1931) 113–129.
(6) "On integral functions of integral order." *Proc. London Math. Soc.* (2) **33** (1932) 209–224.
(7) "On functions which are regular and of finite order in an angle." *Proc. London Math. Soc.* (2) **38** (1935) 158–179.
(8) "On the minimum modulus of integral functions." *Proc. Cambridge Philos. Soc.* **30** (1934) 412–420.
(9) "On certain integral functions of order 1 and mean type." *Proc. Cambridge Philos. Soc.* **31** (1935) 347–350.
(10) "On the directions of Borel of functions which are regular and of finite order in an angle." *Proc. London Math. Soc.* (2) **38** (1935) 503–541.
(11) "Some uniqueness theorems." *Proc. London Math. Soc.* (2) **41** (1936) 33–47.
(12) "On certain integral functions of order 1 " *Quart. J. Math. Oxford Ser.* (1) **7** (1936) 46–55.
(13) "On functions bounded at the lattice points in an angle." *Proc. London Math. Soc.* (2) **43** (1937) 26–32.
(14) "On the maximum modulus principle for functions with zeros and poles." *Proc. London Math. Soc.* (2) **32** (1931) 51–71.
(15) "Some generalizations of Montel's theorem." *Proc. Cambridge Philos. Soc.* **31** (1935) 26–30.
(16) "On the directions of Borel of analytic functions." *Proc. London Math. Soc.* **(2) 38** (1935) 417–457.

CARTWRIGHT, M. L., and COLLINGWOOD, E. F.
(1) "Boundary theorems for a function meromorphic in the unit circle." *Acta Math.* **87** (1952) 83–146.

CESÀRO, E.
(1) "Sur les fonctions holomorphes de genre quelquonque." *C. R. Acad. Sci. Paris* **99** (1884), 26–27.

CHANDRASEKHARAN, K.
(1) "On Hadamard's factorization theorem." *J. Indian Math. Soc.* **5** (1941) 128–132.

CHANG, SHIH-HSUN
(1) "On a theorem of S. Bernstein." *Proc. Cambridge Philos. Soc.* **48** (1952) 87–92.

CHEBOTARËV, N. G., and MEIMAN, N. N.
(1) "The Routh-Hurwitz problem for polynomials and entire functions." Trudy Mat. Inst. Steklov., **26** 1949 (Russian).

CIVIN, P.
(1) "Inequalities for trigonometric integrals." *Duke Math. J.* **8** (1941) 656–665.

CLUNIE, J.
(1) "The minimum modulus of a polynomial on the unit circle." *Quart. J. Math. Oxford Ser.* (2) **10** (1959) 95–98.
(2) "On integral and meromorphic functions." *J. London Math. Soc.* **37** (1962) 17–27.
(3) "On the determination of an integral function of finite order from its Taylor series." *J. London Math. Soc.* **28** (1953) 58–66.
(4) "Note on a theorem of Parthasanathy." *J. London Math. Soc.* **28** (1953) 377–379.
(5) "Univalent regions of integral functions." *Quart. J. Math. Oxford* (2) **5** (1954) 291–296.
(6) "On the determination of an integral function from its Taylor series." *J. London Math. Soc.* **30** (1955), 34–42.
(7) "The asymptotic paths of integral functions of infinite order." *J. London Math. Soc.* **30** (1955) 80–83.
(8) "The asymptotic behaviour of integral functions." *Quart. J. Math. Oxford* (2) **6** (1955) 1–3.
(9) "Note on integral functions of infinite order." *Quart. J. Math. Oxford* (2) **6** (1955) 88–90.
(10) "The maximum modulus of an integral function of an integral function." *Quart. J. Math. Oxford* (2) **6** (1955) 176–178.
(11) "The behaviour of integral functions determined from their Taylor series." *Quart. J. Math. Oxford* (2) **7** (1956) 175–182.
(12) "The derivative of a meromorphic function." *Proc. Amer. Math. Soc.* **7** (1956) 227–229.
(13) "On functions meromorphic in the unit circle." *J. London Math. Soc.* **32** (1957) 65–67.
(14) "On a theorem of Noble." *J. London Math. Soc.* **32** (1957) 138–144.
(15) "On a paper of Kennedy." *J. London Math. Soc.* **33** (1958) 118–120.
(16) "Inequalities for integral functions." *Quart. J. Math. Oxford* (2) **9** (1958) 1–7.
(17) "On the maximum modulus of an integral function." *Proc. London Math. Soc.* (3) **10** (1960) 161–179.

(18) "On integral functions having prescribed growth." *Can. J. Math.* **17** (1965) 396–404.
(19) "On the composition of entire and meromorphic functions." "Macintyre Memorial Volume." Ohio State Univ. Press, Columbus, Ohio, 1971.

CLUNIE, J., and ANDERSON, J. M.
(1) "Slowly growing meromorphic functions." *Comment. Math. Helv.* **40** (1966) 267–280.

CLUNIE, J., and HAYMAN, W. K.
(1) "The maximum term of a power series I." *J. Anal. Math.* **12** (1964) 143–186.
(2) "The maximum term of a power series II." *J. Anal. Math.* **14** (1965) 15–65.
(3) "The spherical derivative of integral and meromorphic functions." *Comment. Math. Helv.* **40** (1966) 117–148.

CLUNIE, J., and KÖVARI, T.
(1) "On integral functions having prescribed asymptotic growth II." *Canad. J. Math.* **20** (1968) 7–20.

COLLINGWOOD, E. F.
(1) "Sur les valeurs exceptionelles des fonctions entières d'ordre fini." *C. R. Acad. Sci Paris* **179** (1924) 1125–1127.
(2) "Exceptional values of meromorphic functions." *Trans. Amer. Math. Soc.* **66** (1949) 308–346.

COPSON, E. T.
(1) "An Introduction to the Theory of Functions of a Complex Variable," corrected ed. Oxford Univ. Press, London and New York, 1960.

CRUM, M. M.
(1) "On the resultant of two functions." *Quart. J. Math. Oxford Ser.* (1) **12** (1941) 108–111.

CSILLAG, P.
(1) "Über ganze Funktionen, welche drei nicht verschwindende Ableitungen besitzen." *Math. Ann.* **110** (1935) 745–752.

DAVIS, P.
(1) "Completeness theorems for sets of differential operators." *Duke Math.* **20** (1953) 345–357.

DELANGE, H.
(1) "Sur les suites de polynomes ou de fonctions entières a zéros réels." *Ann. Sci. École Norm. Sup.* (3) **62** (1945) 115–183.
(2) "Un théorème sur les fonctions entières a zéros réels et négatifs." *J. Math. Pures Appl.* (9) **31** (1952) 55–78.

DENJOY, A.
(1) "L'allure asymptotique des fonctions entières d'ordre fini." *C. R. Acad. Sci. Paris* **242** (1956) 213–218.
(2) "Sur un théorème de Wiman." *C. R. Acad. Sci. Paris* **193** (1931) 828–830; correction in Sur quelques points de la théorie des fonctions, *ibid.* **194** (1932) 44–46.
(3) "Sur les fonctions entières de genre fini." *C. R. Acad. Paris* **145** (1907) 106–108.
(4) "Sur les produits canoniques d'ordre infini." Thèse. *J. Math.* (1910).
(5) "Sur l'intégration de certaines inéquations fonctionnelles." *C. R. Acad. Sci. Paris*, **148** (1909) 981–983.

DIENES, P.
(1) "The Taylor series. An Introduction to the Theory of Functions of a Complex Variable." Oxford Univ. Press, London and New York, 1931.

DINGHAS, A.
(1) "Über einen Satz von Phragmén und Lindelöf." *Math. Z.* **39** (1934) 455–461.
(2) "Über das Phragmén-Lindelöfschen Prinzip und den Julia-Carathéodoryschen Satz." *Sitzungsber. Preuss. Akad. Wiss.* (1936).
(3) "Eine Verallgemeinerung des Picard–Borelschen Satzes." *Math. Z.* **44** (1939).

DUFFIN, R. J., and SCHAEFFER, A. C.
(1) "Some inequalities concerning functions of exponential type." *Bull. Amer. Math. Soc.* **43** (1937) 554–556.
(2) "Some properties of functions of exponential type." *Bull. Amer. Math. Soc.* **44** (1938) 236–240.
(3) "Power series with bounded coefficients." *Amer. J. Math.* **67** (1945) 141–154.
(4) "A class of nonharmonic Fourier series." *Trans. Amer. Math. Soc.* **72** (1952) 341–366.

DUFRESNOY, J.
(1) "Sur les domains couvertes par les valeurs d'une fonction méromorphe ou algébroïde." *Ann. Sci. École Norm. Sup.* (3) **58** (1941) 179–259.
(2) "Sur le produit de composition de deux fonctions." *C. R. Acad. Sci. Paris* **225** (1947) 857–859.
(3) "Sur les fonctions méromorphes et univalentes dans le cercle unité." *Bull. Sci. Math.* **69** (1945) 21–22.

DUFRESNOY, J., and PISOT, C.
(1) "Prolongement analytique de la série de Taylor." *Ann. Sci. École Norm. Sup.* (3) **68** (1951) 105–124.

DUGUÉ, D.
(1) "Le défaut au sens de M. Nevanlinna dépend de l'origine choisie." *C. R. Acad. Sci. Paris* **225** (1947) 555–556.
(2) "Sur certaines conséquences qu'entraîne pour une série de Fourier le fait d'avoir tous ses coefficients positifs. Complément au théorème de Weierstrass." *C. R. Acad. Sci. Paris* **228** (1949) 1469–1470.

DZHRBASHYAN, M. M.
(1) "Uniqueness and representation theorems for entire functions." *Izv. Akad. Nauk SSSR. Ser. Mat.* **16** (1952) 225–252 (Russian).
(2) "On the integral representation and uniqueness of some classes of entire functions." *Mat. Sb.* **33** (75) (1953) 485–530 (Russian).

EDREI, A.
(1) "Meromorphic functions with three radially distributed values." *Trans. Amer. Math. Soc.* **78** (1955) 276–293.
(2) "Sums of deficiencies of meromorphic functions." *J. Anal. Math.* **14** (1965) 79–107.

EDREI, A., and FUCHS, W. H. J.
(1) "On the zeros of $f(g(z))$ where f and g are entire functions." *J. Anal. Math.* **12** (1964) 243–255.
(2) "On the growth of meromorphic functions with several deficient values." *Trans. Amer. Math. Soc.* **93** (1959) 292–328.

(3) "Valeurs déficientes et valeurs asymptotiques des fonctions méromorphes." *Comment. Math. Helv.* **33** (1959) 258–295.
(4) "The deficiencies of meromorphic functions of order less than one." *Duke Math. J.* **27** (1960) 233–249.
(5) "Bounds for the number of deficient values of certain classes of meromorphic functions." *Proc. London Math. Soc.* **12** (1962) 315–344.

EDREI, A., FUCHS, W. H. J., and HELLERSTEIN, S.
(1) "Radial distribution and deficiencies of the values of a meromorphic function." *Pac. J. Math.* **11** (1961) 135–151.

EGOROFF, D. T.
(1) "Sur les suites de fonctions mesurables." *C. R. Acad. Sci. Paris* **152** (1911) 244–246.

EIDSWICK, J. A.
(1) "A hereditary class of Hilbert spaces of entire functions." Dissertation, Purdue University, Lafayette, Indiana, 1964.

ESSEN, M.
(1) Note on "A theorem on the minimum modulus of entire functions by Kjellberg." *Math. Scand.* **12** (1963) 12–14.

EWEIDA, M. T.
(1) "A note on the generalization of Taylor's expansion." *Proc. Math. Phys. Soc. Egypt* **3** (1946) No. 2 (1947) 1–7.
(2) "On the representation of integral functions by generalized Taylor's series." *Proc. Math. Phys. Soc. Egypt* **3** No. 4 (1948) 39–46.

FABER, G.
(1) "Über das Anwachsen analytischer Funktionen." *Math. Ann.* **63** (1907) 549–551.
(2) "Beitrage zur Theorie der ganzen Funktionen." *Math. Ann.* **70** (1911) 48–68.

FATOU, P.
(1) "Sur les équations fonctionelles." *Bull. Soc. Math. France* **47** (1919), 161–271; **48** (1920) 33–94, 208–314.
(2) "Sur l'itération des fonctions transcendantes entières." *Acta Math.* **47** (1926) 337–370.
(3) "Séries trigonométriques et séries de Taylor." *Acta Math.* **30** (1906) 335–400.

FEJÉR, L., and RIESZ, F.
(1) "Über einige funktionentheoretische Ungleichungen." *Math. Z.* **11** (1921) 305–314.

FELLER, W.
(1) "A simple proof for renewal theorems." *Comm. Pure Appl. Math.* **14** (1961) 285–293.

FLETT, T. M.
(1) "Note on a function-theoretic identity." *J. London Math. Soc.* **29** (1954) 115–118.

FRANCK, A.
(1) "Analytic functions of bounded type." *Amer. J. Math.* **74** (1952) 410–422.

FROSTMAN, C.
(1) "Über die defekten Werte einer meromorphen Funktion." *C. R. 8^e Congr. Math. Scand. Stockholm* (1934) 392–396.

FUCHS, W. H. J.
(1) "Proof of a conjecture of G. Pólya concerning gap series." *Illinois J. Math.* **7** (1963) 661–667.
(2) "A theorem on the Nevanlinna deficiencies of meromorphic functions of finite order." *Ann. Math.* (2) **68** (1958) 203–209.
(3) "A generalization of Carlson's theorem." *J. London Math. Soc.* **21** (1946) 106–110.

GABRIEL, R. M.
(1) "Some results concerning the integrals of moduli of regular functions along certain curves." *J. London Math. Soc.* **2** (1927) 112–117.

GAIER, D.
(1) "Zur Frage der Indexverschiebung beim Borel-Verfahren." *Math. Z.* **58** (1953) 453–455.

GANAPATHY, I. V.
(1) "On permutable integral functions." *J. London Math. Soc.* **34** (1959) 141–144.
(2) "On the Lebesgue class of integral functions along straight lines issuing from the origin." *Quart. J. Math. Oxford Ser.* (1) **7** (1936) 294–299.
(3) "On the order and type of integral functions bounded at a sequence of points." *Ann. Math.* (2) **38** (1937) 311–320.
(4) "On effective sets of points in relation to integral functions." *Trans. Amer. Math. Soc.* **42** (1937), 358–365; correction, *ibid.* **43** (1938) 494.
(5) "A note on integral functions of order one." *Quart. J. Math. Oxford Ser.* (1) **8** (1937) 103–106.
(6) "A note on integral functions of order 2 bounded at the lattice points." *J. London Math. Soc.* **11** (1936) 247–249.
(7) "Some properties of integral functions of finite order." *Quart. J. Math. Oxford Ser.* (1) **8** (1937) 131–141.
(8) "On integral functions of finite order and minimal type." *J. Indian Math. Soc.* **2** (1937) 131–140.
(9) "A property of the zeros of the successive derivatives of integral functions." *J. Indian Math. Soc.* **2** (1937) 289–294.
(10) "On the average radial increase of a certain class of integral functions of order one and finite type." *J. Indian Math. Soc.* **3** (1938) 87–95.
(11) "Some theorems of functions regular in an angle." *Quart. J. Math. Oxford Ser.* (1) **9** (1938) 206–215.
(12) "The behaviour of integral functions at the lattice-points." *J. London Math. Soc.* **13** (1938) 91–94.
(13) "The Phragmén-Lindelöf theorem in the critical angle." *J. London Math. Soc.* **14** (1939) 286–292.
(14) "The influence of zeros on the magnitude of functions regular in an angle." *J. Indian Math. Soc.* **7** (1943) 1–16.

GEL'FOND, A. O.
(1) "Sur une application du calcul des différences finies a l'étude des fonctions entières." *Rec. Math.* (*Mat. Sb.*) **36** (1929) 173–183.
(2) "Interpolation et unicité des fonctions entières." *Rec. Math.* (*Mat. Sb.*) **4** (46) (1938) 115–147.
(3) "On the Taylor series associated with an integral function." *C. R.* (*Dokl.*) *Acad. Sci. URSS* **23** (1939) 756–758.

(4) "Linear differential equations of infinite order with constant coefficients and asymptotic periods of entire functions." *Tr. Mat. Inst. Steklov.*, **38** (1951) 42–67 (Russian); translated as *Amer. Math. Soc. Transl.* No. 84 (1953).

GEL'FOND, A. O., and IBRAGIMOV, I. I.

(1) "On functions whose derivatives are zero at two points." *Izv. Akad. Nauk SSSR. Ser. Mat.* **11** (1947) 547–560 (Russian).

GERMAY, R. H. J.

(1) "Sur une application des théorèmes de Weierstrass et de Mittag–Leffler de la théorie générale des fonctions." *Ann. Soc. Sci. Bruxelles Ser. I* **60** (1946) 190–195.

GIACCARDI, F.

(1) "Su di una condizione perché una funzione analitica periodica si riduca ad un polinomio trigonometrico." *Atti Accad. Naz. Lincei. Rend. Cl. Sci. Fis. Mat. Natur.* (6) **25** (1937) 555–559.

GOL'DBERG, A.A.

(1) "On the possible value of the lower order of an entire function with a finite deficient value." *Dokl. Akad. Nauk SSSR* **159** (1964) 968–970 (Russian).

(2) "On the inverse problem of the theory of the distribution of values of meromorphic functions." *Ukrain. Mat. Zh.* **6** (1954) 385–397. (Russian)

(3) "On the deficiencies of meromorphic functions." *Dokl. Akad. Nauk SSSR* **98** (1954) 893–895 (Russian).

(4) "On an inequality for logarithmically convex functions" *Dopovidi Akad. Nauk Ukrain. R.S.R.* (1957) 227–230 (Ukrainian).

(5) "On the set of deficient values of meromorphic functions of finite order." *Ukrain. Mat. Zh.* **11** (1959) 438–443 (Russian).

GOL'DBERG, A. A., and OSTROVSKIJ, I. V.

(1) "New investigations on the growth and distribution of values of entire and meromorphic functions of genus zero." *Usp. Mat. Nauk* **16** (1961) 51–62.

GOL'DBERG, A. A., and TAIROVA, V. G.

(1) "On integral functions with two finite completely multiple values." *Zap. Mech.-Mat.-Fak. Harkov. Gos. Univ.* **29** (1963), 67–78 (Russian).

GONTCHAROFF, W.

(1) "Détermination des fonctions entières par interpolation." "Actualités Scientifiques et Industrielles," No. 465. Hermann, Paris, 1937.

GOODMAN, A. W., RAHMAN, Q. I., and RATTI, J. S.

(1) "On the zeros of a polynomial and its derivative." *Proc. Amer. Math. Soc.* **21** (2) (1969) 273–274.

GRANDJOT, K.

(1) "Über Polynome, die in Einheitswurzeln beschränkt sind." *Jber. Deutsch. Math. Verein.* **34** (1925) 80–86; correction, *ibid.* **35** (1926) 112.

GRONWALL, T. H.

(1) "A sequence of polynomials connected with the *n*th roots of unity." *Bull. Amer. Math. Soc.* **27** (1921) 275–279.

GROSS, F.

(1) "On the periodicity of compositions of entire functions." *Canad. J. Math.* **18** (1966) 724–730.

(2) "Entire solutions of the functional equation $\alpha(\beta(z)) = \alpha(\gamma(z)) + c$." *J. Indian Math. Soc.* **32** (1968) 199–206.
(3) "On factorization of meromorphic functions." Lecture given at the Mathematical Institute on entire functions and related topics, Univ. of California, La Jolla, California, June 28 (1966).

GROSS, W.
(1) "Eine ganze Funktion für die jede komplexe Zahl Konvergenzwert ist." *Math. Ann.* **79** (1918) 201–208.

GURIN, L. S.
(1) "On an interpolation problem." *Mat. Sb.* **22** (64) (1948) 425–438 (Russian).

HADAMARD, J.
(1) "Sur les propriétés des fonctions entières et en particulier une fonction étudiée par Riemann." *J. Math.* (4) **9** (1893) 171–215.
(2) "Essai sur l'étude des fonctions données par leur développment de Taylor." *J. Math.* (4) **8** (1892) 101–185.
(3) "Sur la croissance des fonctions entières." *Bull. Soc. Math.* 1896.
(4) "Sur les fonctions entières." *C. R. Acad. Sci. Paris* **135** (1902) 1309–1311.
(5) "Théorème sur les séries entières." *Acta Math.* **22** (1899) 55–64.

HALLSTROM, G. F.
(1) "Über meromorphen Funktionen mit mehrfach zusammenhängender Existenzgebieten." *Acta Acad. Abo. Ser. B* **12** (1940).

HARDY, G. H.
(1) "Divergent Series." Oxford Univ. Press, London and New York, 1949.
(2) "The maximum modulus of an integral function." *Quart. J.* **41** (1910) 1–9.
(3) "The mean value of the modulus of an analytic function." *Proc. London Math. Soc.* (2) **14** (1915) 269–277.

HARVEY, A. R.
(1) "The mean of a function of exponential type." *Amer. J. Math.* **70** (1948) 181–202.

HAYMAN, W. K.
(1) *Proc. Cambridge Philos. Soc.* **48** (1952) 93–105.
(2) "The growth of entire and subharmonic functions." "Lectures on functions of a Complex Variable." Ann Arbor, Michigan, 1955, 182–198. *Math. Rev.* **23** (1962) A.3264 (*et al.*).
(3) "Some applications of the transfinite diameter to the theory of functions." *J. Anal. Math.* **1** (1951) 155–179.
(4) "Uniformly normal families." "Lectures on functions of a Complex Variable." Univ. Michigan Press, Ann Arbor, 1955, pp. 199–212.
(5) "Multivalent Functions." Cambridge Univ. Press, London and New York, 1958.
(6) "Slowly growing integral and subharmonic functions." *Comment. Math. Helv.* **34** (1960) 75–84.
(7) "On functions with positive real part." *J. London Math. Soc.* **36** (1961) 35–48.
(8) "On the growth of integral functions on asymptotic paths." *J. Indian Math. Soc.* **24** (1960) 251–264.
(9) "On the limits of moduli of analytic functions." *Ann. Polon. Math.* **12** (1962) 143–150.

(10) "Meromorphic Functions," Oxford Mathematical Monograph. Oxford Univ. Press (Clarendon), London and New York, 1964.

(11) "Some integral functions of infinite order." *Math. Notae. Bol. Inst. Mat.* **20** (1965) 1–5.

(12) "Die Nevanlinna-Charakteristik von meromorphen Funktionen und ihren Integralen." "Festband zum 70. Geburtstag von Rolf Nevanlinna." Springer-Verlag, Berlin, 1966, pp. 16–20.

(13) "Research Problems in Function Theory." Oxford Univ. Press (Athlone), London and New York, 1967.

(14) "On direct critical singularities and regularity of growth." *J. Anal. Math.* **18** (1967) 113–120.

(15) "Regular Tsuji functions with infinitely many Julia points." *Nagoya Math. J.* **29** (1967) 185–196.

(16) "Note on Hadamard's convexity theorem." *Proc. Symp. Pure Math.* Volume XI, "Entire Functions and Related Parts of Analysis." Amer. Math. Soc., Providence, Rhode Island, 1968, pp. 210–213.

(17) "On integral functions with distinct asymptotic values." *Proc. Cambridge Philos. Soc.* **66** (1969) 301–315.

(18) "The Minimum modulus of large integral functions." *Proc. London Math. Soc.* (3) **2** (1952) 469–512.

(19) "Some remarks on Schottky's theorem." *Proc. Cambridge Philos. Soc.* **43** (1947) 442–454.

(20) "Some inequalities in the theory of functions." *Proc. Cambridge Philos. Soc.* **44** (1948) 159–178.

(21) "Sur le module des fonctions entières." *C. R. Acad. Sci. Paris* **232** (1951) 591–593.

(22) "A characterisation of the maximum modulus of functions regular at the origin." *J. Anal. Math.* **1** (1951) 135–154.

(23) "The maximum modulus and valency of functions meromorphic in the unit circle." *Acta Math.* **86** (1951) 88–257.

(24) "Functions with values in a given domain." *Proc. Amer. Math. Soc.* **3** (1952) 428–432.

(25) "An integral function with a defective value, that is neither asymptotic nor invariant under change of origin." *J. London Math. Soc.* **28** (1953) 369–376.

(26) "On Nevanlinna's second theorem and extensions." *Rend. Circ. Mat. Palermo* (2) **2** (1953) 346–392.

(27) "A generalisation of Stirling's formula." *J. Reine Angew. Math.* **196** (1956) 67–95.

(28) "Picard values of meromorphic functions and their derivatives." *Ann. Math.* **70** (1959) 9–42.

(29) "On the characteristic of functions meromorphic in the unit disk and of their integrals." *Acta Math.* **112** (1964) 181–214.

(30) "On the characteristic of functions meromorphic in the plane and of their integrals." *Proc. London Math. Soc.* (3) **14A**(1965) 93–128.

HAYMAN, W. K., and FUCHS, W. H. J.

(1) "An entire function with assigned deficiencies," *in* "Studies in Mathematical Analysis and Related Topics." Stanford Univ. Press, Stanford, California, 1962, pp. 117–125.

HAYMAN, W. K., and STEWART, F. M.

(1) "Real inequalities with applications to function theory." *Proc. Cambridge Philos. Soc.* **50** (1954) 250–260.

HEINS, M. H.

(1) "Entire functions with bounded minimum modulus; subharmonic functions analogues." *Ann. Math.* (2) **49** (1948) 200–213.

(2) "On the Phragmén-Lindelöf principle." *Trans. Amer. Math. Soc.* **60** (1946) 238–244.

(3) "On some theorems associated with the Phragmén-Lindelöf principle." *Ann. Acad. Sci. Fenn. Ser. A I* **46** (1948) 1–10.

(4) "On the Denjoy–Carleman–Ahlfors theorem." *Ann. of Math.* (2) **49** (1948) 533–537.

HERVE, M.

(1) "Sur quelques applications de la notion d'ordere précisé." *Bull. Sci. Math.* (2) **66** (1942) 17–24, 31–48.

HILLE, E.

(1) "Analytic Function Theory," Vol. 2. (Ginn Blaisdell) Boston Massachusetts, 1962.

(2) "Analytic Function Theory," Vol. 1. (Ginn Blaisdell) Boston, Massachusetts, 1959, pp. 225–229.

HOHEISEL, G.

(1) "Über das Verhalten einer analytischen Funktion in einer Teilumgebung eines singulären Punktes." *Sitzungsber Preuss. Akad. Wiss. Phys.-Math. Kl.* (1923) 177–180.

HOPF, E.

(1) "Fuchsian groups and ergodic theory." *Trans. Amer. Math. Soc.* **39** (1936).

HUBER, A.

(1) "Über Wachstumseigenschaften gewisser Klassen von subharmonischen Funktionen," "Dissertation, Zurich, 1951; also *Comm. Math. Helv.* **26** (1952) 81–116.

HURWITZ, A.

(1) "Über die Theorie der elliptischen Modulfunktion." *Math. Ann.* **58** (1904) 343–360.

(2) "Über die Anwendung der elliptischen Modulfunktion auf einen Satz der allgemeinen Funktionentheorie." Viertelj. der Naturf. Gesellschaft., Zürich, 1904.

(3) "Sur les points critiques des fonctions inverses." *C. R. Acad. Sci. Paris,* **143** (1906) 877–879.

(4) "Sur les points critiques des fonctions inverses." *C. R. Acad. Sci. Paris,* **144** (1907) 63–65.

HYLLENGVEN, A.

(1) "Valiron deficient values for meromorphic functions in the plane." *Acta. Math.* **124**, (1970) 1–8.

IBRAGIMOV, I. I.

(1) "Sur quelques systèmes complets de fonctions analytiques." *Izv. Akad. Nauk SSSR Sér. Mat.* **3** (1939) 553–568 (Russian; French summary.)

IKEHARA, S.

(1) "On integral functions with real negative zeros." *J. Math. Phys.* **10** (1931) 84–91.

INOUE, M.

(1) "Sur le module minimum des fonctions sousharmoniques et des fonctions entières d'ordre $<\frac{1}{2}$." *Mem. Fac. Sci. Kyūsyū Univ. A.* **4** (1949) 183–193.

(2) "On the growth of subharmonic functions and its applications to a study of the minimum modulus of integral functions." *J. Inst. Polytech. Osaka City Univ. Ser. A.* **1**, No. 2 (1950) 71–82.
(3) "A note on minimum modulus of integral functions of lower order $<\frac{1}{2}$." *Math. Japon.* **2** (1950) 41–47.

IVERSEN, F.
(1) "Recherches sur les fonctions inverses des fonctions méromorphes." Thesis, Univ. of Helsinki, 1914.
(2) "Sur une fonction entière dont la fonction inverse présente un ensemble de singularités de la puisance du continu." *Ofvers. Finska Soc.* **58**.
(3) "Sur quelques propriétés des fonctions monogènes au voisinage d'un point singulier." *Ofvers. Finska Soc.* **58**.
(4) "Sur quelques fonctions entières qui admettent des valeurs asymptotiques finies." *Ofvers. Finska Soc.* **61**.
(5) "Zum Verhalten analytischer Funktionen in Bereichen, deren Rand eine wesentliche singularität enthält." *Ofvers. Finska Soc.* **64**.
(6) "Sur les valeurs asymptotiques des fonctions méromorphes et les singularités transcendantes de leurs inverses." *C. R. Acad. Sci. Paris* **166** (1918) 156–159.

JACOBSTHAL, E.
(1) "Über vertauschbare Funktionen." *Math. Z.* **63** (1955) 243–276.

JAIN, S. P.
(1) "An analogue of a theorem of Phragmén-Lindelöf." *J. Indian Math. Soc.* **19** (1932) 241–245.

JENKINS, J. A.
(1) "On explicit bounds in Schottky's theorem." *Canad. J. Math.* **7** (1955) 76–82.
(2) "On explicit bounds in Landau's theorem." *Canad. J. Math.* **8** (1956) 423–425.

JENSEN, J. L. W. V.
(1) "Sur un nouvel et important théorème de la théorie des fonctions." *Acta Math.* **22** (1899) 359–364.

JULIA, G.
(1) "Mémoire sur l'itération des fonctions rationelles." *J. Math. Pures Appl.* (8) **1** (1918) 47–245.
(2) "Mémoire sur la permutabilité des fonctions rationelles." *Ann. Sci. École Norm. Sup.* (3) **39** (1922) 131–215.
(3) "Leçons sur les fonctions uniformes a point singulier essentiel isolé." Borel Monograph, Gauthier-Villars Paris (1923).
(4) "Sur quelques propriétés nouvelles des fonctions entières ou méromorphes, 1st mémoire." *Ann. Sci. École Norm. Sup.* **36** (1919).
(5) "2nd mémoire." *Ann. Sci. École Norm. Sup.* **37** (1919).
(6) "3rd mémoire." *Ann. Sci. École Norm. Sup.* **38** (1921).

JUNNILA, A.
(1) "Über das Anwachsen einer analytischen Funktion in einer gegebenen Punktfolge." *Ann. Acad. Sci. Fenn. Ser. A* **47** (1936) No. 2.

KAHANE, J. P.
(1) "Extension du théorème de Carlson et applications." *C. R. Acad. Sci. Paris* **234** (1952) 2038–2040.

KAMETANI, S.

(1) "The exceptional values of functions with the set of capacity zero of essential singularities." *Proc. Imp. Acad. Japan* **17** (1941).

(2) "On Hausdorff's measures and generalized capacities with some of their applications to the theory of functions." *Japan. J. Math.* **19** (1944–48).

KAWATA, T.

(1) "Remarks on the representation of entire functions of exponential type." *Proc. Imp. Acad. Japan* **14** (1938) 266–269.

KENNEDY, P. B.

(1) "A class of integral functions bounded on certain curves." *Proc. London Math. Soc.* (3) **6** (1956) 518–547.

(2) "On a Conjecture of Heins." *Proc. London Math. Soc.* (3) **5** (1955) 22–47.

KENŽEGULOV, H. K.

(1) "On certain questions in the theory of iteration of entire functions." *Volž. Mat. Sb. Vyp.* **2** (1964) 56–60 (Russian).

KJELLBERG, B.

(1) "On certain integral and harmonic functions." Thesis, Uppsala, 1948.

(2) "On the minimum modulus of entire functions of lower order less than one." *Math. Scand.* **8** (1960) 189–197.

(3) "A theorem on the minimum modulus of entire functions." *Math. Scand.* **12** (1963) 5–11.

(4) "A relation between the maximum and minimum modulus of a class of entire functions." C. R. du 12. Congrès des Mathématiciens Scandinaves tenu à Lund 10–15 août 1953, Lund. 1954, 135–138.

(5) "On integral functions bounded on a given set." *Mat. Tidsskrift B* (1952) 92–99.

KNOPP, K.

(1) "Theory and Application of Infinite Series." Blackie, London and Glasgow, 1928 (2nd Ed. 1951).

KOBER, H.

(1) "On the approximation to integrable functions by integral functions." *Trans. Amer. Math. Soc.* **54** (1943) 70–82.

(2) "Approximation of continuous functions by integral functions of finite order." *Trans. Amer. Math. Soc.* **61** (1947) 293–306.

KOMATU, Y.

(1) "The order of the derivative of a meromorphic function." *Proc. Japan Acad.* **27** (1951).

KOREVAAR, J.

(1) "An inequality for entire functions of exponential type " *Nieuw Arch. Wiskunde* (2) **23** (1949) 55–62.

(2) "A simple proof of a theorem of Pólya." *Simon Stevin* **26** (1949) 81–89.

(3) "Approximation and interpolation applied to entire functions." Thesis, Univ. of Leiden, 1949. The second part appeared also as "Functions of exponential type bounded on sequences of points." *Ann. Soc. Polon. Math.* **22** (1949, 1950) 207–234.

(4) "The zeros of approximating polynomials and the canonical representation of an entire function." *Duke Math. J.* **18** (1951) 573–592.

KÖVARI, T.

(1) "On the asymptotic paths of entire functions with gap power series." *J. Anal. Math.* **15** (1965) 281–282.
(2) "A gap theorem for entire functions of infinite order." *Michigan Math. J.* **12** (1965) 133–140.
(3) "Asymptotic values of entire functions of finite order with density conditions." *Acta Sci. Math. Szeged* **26** (1965) 233–237.
(4) "The growth of entire functions of finite order with density conditions." *Quart. J. Math. Oxford Ser.* (2) **17** (1966) 22–30.

KRAFT, A.

(1) Über ganze transzendente Funktionen von unendlicher Ordnung." Inaugural dissertation, Univ. of Göttingen, 1903.

KREIN, M. G.

(1) "A contribution to the theory of entire functions of exponential type." *Izv. Akad. Nauk SSSR Sér. Mat.* **11** (1947) 309–326 (Russian; English summary).

KUNUGUI, K.

(1) "Une géneralisation des théorèmes de MM. Picard-Nevanlinna sur les fonctions méromorphes." *Proc. Imp. Acad. Japan* **17** (1941).
(2) "Sur la théorie des fonctions méromorphes et uniforme." *Japan. J. Math.* **18** (1941–1943).

LAI, W. T.

(1) "Über den Satz von Landau." *Sci. Rec.* **4** (1960) 339–342.

LAKSHMINARASIMHAN, T. V.

(1) "A Tauberian theorem for the type of an entire function." *J. Indian Math. Soc.* **17** (1953) 55–58.

LANDAU, E.

(1) "Über eine Verallgemeinerung des Picardschen Satzes." *Sitzungsber. Preuss. Akad. Wiss. Phys.-Math. Kl.* (1904) 1118–1133.
(2) "Handbuch der Lehre von der Verteilung der Primzahlen." Teubner, Leipzig and Berlin, 1909.
(3) "Vorlesungen über Zahlentheorie." Hirzel, Leipzig, 1927.
(4) "Darstellung und Begründung einiger neuerer Ergebnisse der Funktionen-theorie," 2nd ed. Springer-Verlag, Berlin and New York, 1929.
(5) "Über der Picardschen Satz." *Vierteljahresschr. Naturforsch. Ges. Zürich* (1906).
(6) "Zum Koebeschen Verzerrungssatz." *Rend. Circ. Mat. Palermo* **46** (1922) 347–348.

LANDAU, E., and VALIRON, G.

(1) "A deduction from Schwarz's lemma." *J. London Math. Soc.* **4** (1929).

LEHTO, O.

(1) "The spherical derivative of a meromorphic function in the neighbourhood of an isolated singularity." *Comment. Math. Helv.* **33** (1959) 196–205.

LEHTO, O., and VIRTANEN, K. I.

(1) "Boundary behaviour and normal meromorphic functions." *Acta Math.* **97** (1957) 47–65.

LELONG-FERRAND, J.

(1) Étude au voisinage de la frontière des fonctions surharmoniques positives dans un demi-espace." *Ann. Sci. École Norm. Sup.* (3) **66** (1949) 125–159.

LE-VAN THIEM

(1) Über das Umkehrproblem der Wertverteilungslehre." *Comment. Math. Helv.* **23** (1949) 26–49.

LEVIN, B. YA.

(1) "Sur la croissance d'une fonction entière suivant un rayon et la distribution de ses zéros suivant leurs arguments." *Mat. Sb.* **2** (44) (1937) 1097–1142 (Russian; French summary).

(2) "Sur certaines applications de la série d'interpolation de Lagrange dans la théorie des fonctions entières." *Mat. Sb.* **8** (50) (1940) 437–454 (Russian; French summary).

(3) "On functions determined by their values on an interval." *Dokl. Akad. Nauk SSSR* **70** (1950) 757–760 (Russian).

(4) "On entire functions of finite degree which are of regular growth." *Dokl. Akad. Nauk SSSR* **71** (1950) 601–604 (Russian).

(5) "On a special class of entire functions and on related extremal properties of entire functions of finite degree." *Izv. Akad. Nauk SSSR. Ser. Mat.* **14** (1950) 45–84 (Russian).

(6) "On a class of entire functions." *Dokl. Akad. Nauk SSSR* **78** (1951) 1085–1088 (Russian).

(7) "The general form of special operators on entire functions of finite degree." *Dokl. Akad. Nauk SSSR* **79** (1951) 397–400 (Russian).

(8) "Distribution of zeros of Entire functions." *A.M.S. Trans. of Math. Monograph* No. 5 (1964) viii, 493 pp.

LEVINSON, N.

(1) "On a problem of Pólya." *Amer. J. Math.* **58** (1936) 791–798.

(2) "On certain theorems of Pólya and Bernstein." *Bull. Amer. Math. Soc.* **42** (1936) 702–706.

(3) "On the growth of analytic functions." *Trans. Amer. Math. Soc.* **43** (1938) 240–257.

(4) "Gap and Density Theorems." Amer. Math. Soc., Providence, Rhode Island, 1940.

(5) "A theorem of Boas." *Duke Math. J.* **8** (1941) 181–182.

(6) "An integral inequality of the Phragmén-Lindelöf type." *J. Math. Phys.* **20** (1941) 89–98.

LINDELÖF, E.

(1) "Mémoire sur la théorie des fonctions entières de genre fini." *Acta Soc. Sci. Fenn.* **31** No. 1 (1902).

(2) "Sur les fonctions entières d'ordre entier." *Ann. Sci. École Norm. Sup.* (3) **22** (1905) 369–395.

(3) "Sur la détermination de la croissance des fonctions entières définies par un développement de Taylor." *Bull. Sci. Math.* **27** (1903) 213–232.

(4) "Sur un théorème de M. Hadamard dans la théorie des fonctions entières." *Rend. Circ. Mat. Palermo*, **25**.

(5) "Mémoire sur certaines inégalités dans la théorie des fonctions monogènes." *Acta Soc. Sci. Fenn.* **35**.

(6) "Sur un principe général de l'analyse et ses applications a la théorie de la représentation conforme." *Acta Soc. Sci. Fenn.* **46** (1915) 1–35.

LINDWART, E.

(1) "Über eine Methode von Laguerre zur Bestimmung des Geschlechts einer ganzen Funktion." Diss. Inaug., Göttingen, 1914.

LITTLEWOOD, J. E.

(1) "Lectures on the theory of Functions." Oxford Univ. Press, London and New York, 1944.
(2) "On some conjectural inequalities with applications to the theory of integral functions." *J. London Math. Soc.* **27** (1952) 387–393.
(3) "On the asymptotic approximation of integral functions of zero order." *Proc. London Math. Soc.* 2nd ser. **5** (1907) 361–410.
(4) "A general theorem on integral functions of finite order." *Proc. London Math. Soc.* (2) **6** (1908) 189–204.
(5) "On the Dirichlet series and asymptotic expansions of integral functions of zero order." *Proc. London Math. Soc.* **7** (1909) 209–262.

LOHIN, I. F.

(1) "Concerning the representation of entire analytic functions." *Dokl. Akad. Nauk SSSR* **66** (1949) 157–160 (Russian).
(2) "On a representation of an entire function of the first order of normal type." *Dokl. Akad. Nauk SSSR* **72** (1950) 629–632 (Russian).

MACINTYRE, A. J.

(1) "Asymptotic paths of integral functions with gap power series." *Proc. London Math. Soc.* (3) **2** (1952) 286–296.
(2) "Laplace's transformation and integral functions." *Proc. London Math. Soc.* (2) **45** (1938) 1–20.
(3) "Wiman's method and the 'flat regions' of integral functions." *Quart. J. Math. Oxford Ser.* (1) **9** (1938) 81–88.
(4) "The minimum modulus of integral functions of finite order." *Quart. J. Oxford Ser.* (1) **9** (1938) 182–184.
(5) "Interpolation series for integral functions of exponential type." *Trans. Amer. Math. Soc.* **76** (1954) 1–13.
(6) "On the asymptotic paths of integral functions of finite order." *J. London Math. Soc.* **10** (1935) 34–39.

MACINTYRE, S. S.

(1) "On the asymptotic periods of integral functions." *Proc. Cambridge Philos. Soc.* **31** (1935) 543–554.
(2) "An upper bound for the Whittaker constant W " *J. London Math. Soc.* **22** (1947) (1948) 305–311.
(3) "On the zeros of successive derivatives of integral functions." *Trans. Amer. Math. Soc.* **67** (1949) 241–251.
(4) "An interpolation series for integral functions." *Proc. Edinburgh Math. Soc.* (2) **9** (1953) 1–6.

MACLANE, G. R.

(1) "Asymptotic values of holomorphic functions." *Rice Univ. Stud.* **49** (1963) 83.

MACPHAIL, M. S.

(1) "Entire functions bounded on a set." *Trans. Roy. Soc. Canada Sect. III* **37** (1943) 313–338.

MAILLET, E.

(1) "Sur les fonctions entières et quasi-entières." *J. Math.*, 1902.

(2) "Sur les fonctions entières et quasi-entières a croissance régulière." *Ann. Fac. Sci. Univ. Toulouse Ser.* 2, **4**.

(3) "Sur les zéros des fonctions entières, etc." *Ann. Sci. École Norm. Sup.* (1906).

(4) "Sur les fonctions quasi-entières et quasi-méromorphes d'ordre infini non transfini." *Ann. Fac. Sci. Univ. Toulouse Ser.* 2, **9**.

(5) "Sur les fonctions entières d'ordere fini." *Ann. Fac. Sci. Univ. Toulouse Ser.* 2, **10**.

MAITLAND, B. J.

(1) "The flat regions of integral functions of finite order." *Quart. J. Math. Oxford Ser.* (1) **15** (1944) 84–96.

MANDELBROJT, S., and ULRICH, F. E.

(1) "Regions of flatness for analytic functions and their derivatives." *Duke Math. J.* **18** (1951) 549–556.

MARDEN, M.

(1) "On the derivative of an Entire function." *Proc. Amer. Math. Soc.* **74** (1968).

(2) "On the Derivative of Canonical Products." *Pacific J. Math.* **24** (1968) 331–339.

(3) "On the Zeros of the Derivative of an Entire Function." *Amer. Math. Mon.* **75** (1968) 829–839.

(4) "On the Derivative of an Entire Function." *Proc. Amer. Math. Soc.* **19** (1968) 1045–1051.

(5) "On the zeros of rational functions having prescribed poles, with application to the derivative of an Entire Function of finite genre." *Trans. Amer. Math. Soc.* **66** (1949) 407–418.

(6) "Kakeya's problem on the zeros of the derivative of a polynomial." *Trans. Amer. Math. Soc.* **45** (1939) 335–368.

(7) "The geometry of the zeros of a polynomial." Math. Surveys, No. 3, Amer. Math. Soc., Providence, Rhode Island, 1949 (Revised Ed. 1966.)

MARKUSHEVICH, A. I.

(1) "Entire Functions." Amer. Elsevier, New York, 1966.

MARTIN, W. T.

(1) "On expansions in terms of a certain general class of functions." *Amer. J. Math.* **58** (1936) 407–420.

MARTY, F.

(1) "Recherches sur la répartition des valeurs d'une fonction méromorphe." *Ann. Fac. Sci. Univ. Toulouse* (3) **23** (1931) 183–261.

MATTSON, R.

(1) "Sur les fonctions entières d'ordre zéro." Thesis, Univ. of Upsala (1905).

MAZURKIEWICZ, S.

(1) "Sur le terme maximum d'une fonction entière." *C. R. Soc. Sci. Varsovie. Cl. III* **29** (1936) 1–6.

MEIMAN, N. N.
(1) "Differential inequalities and some questions of the distribution of zeros of entire and single-valued analytic functions." *Usp. Mat. Nauk* **7** 3 (49) (1952) 3–62 (Russian).

MIKUSIŃSKI, J. G.
(1) "On the Paley–Wiener theorem." *Stud. Math.* **13** (1953) 287–295.

MILLOUX, H.
(1) "Les fonctions méromorphes et leurs dérivées." Hermann Paris, 1940.
(2) "Les cercles de remplissage des fonctions méromorphes ou entières et le théorème de Picard-Borel." *Acta Math.* **52** (1928) 189–255.
(3) "Sur les directions de Borel des fonctions entières, de leurs derivées et de leurs intégrales." *J. Anal. Math.* **1** (1951) 244–330.

MONTEL, P.
(1) "Leçons sur les familles normales de fonctions analytiques et leur applications." Gauthier-Villars, Paris, 1932.
(2) "Leçons sur les séries de polynômes à une variable complexe." Gauthier-Villars, Paris, 1910.
(3) "Sur les familles de fonctions analytiques qui admettent des valeurs exceptionnelles dans un domaine." *Ann. Sci. École Norm. Sup.* (1912).
(4) "Sur les familles normales de fonctions analytiques." *Ann. Sci. École Norm. Sup.* (1916).
(5) "Sur les fonctions entières de genre fini." *Bull. Sci. Math.* (1922).
(6) "Sur la représentation conforme." *J. Math.* (1917).

MUGGLI, H.
(1) "Differentialgleichungen unendlich hoher Ordnung mit konstanten Koeffizienten." *Comment. Math. Helv.* **11** (1938) 151–179.
(2) "Differentialgleichungen unendlich hoher Ordnung." *Comment. Math. Helv.* **14** (1942) 381–393.

MYRBERG, P. J.
(1) "Eine Verallgemeinerung des Arithmetisch geometrischen Mittels." *Ann. Acad. Sci. Fenn. Ser. A I* **253** (1958) 1–19.
(2) "Iteration von Quadratwurzeloperatoren." *Ann. Acad. Sci. Fenn. Ser. A I* **259** (1958) 1–16.

NEVANLINNA, F.
(1) "Über eine Klasse meromorpher Funktionen." *C. R. 7ᵉ Congr. Math. Scand. Oslo* (1929) 81–83.

NEVANLINNA, F., and NEVANLINNA, R.
(1) "Über die Eigenschaften analytischer Funktionen in der Umgebung einer singulären Stelle oder Linie." *Acta Soc. Sci. Fenn.* **50** (1925).

NEVANLINNA, R.
(1) "Eindeutige analytische Funktionen." Springer-Verlag Berlin, 1953.
(2) "Le théorème de Picard-Borel et la théorie des fonctions méromorphes." Gauthier-Villars, Paris, 1929.
(3) "Über Riemannsche Flächen mit endlich vielen Windungspunkten." *Acta Math.* **58** (1932) 295–373.

(4) "Über die Eigenschaften meromorpher Funktionen in einem Winkelraum." *Acta Soc. Sci. Fenn.* **50** (1925).
(5) "Zur Theorie der meromorphen Funktionen." *Acta Math.* **46** (1925).

NIKOLSKII, S. M.
(1) "Generalization of an inequality of S. N. Bernstein." *Dokl. Akad. Nauk SSSR* **60** (1948) 1507–1510 (Russian).
(2) "Inequalities for entire functions of finite degree and their application in the theory of differentiable functions of several variables." *Tr. Math. Inst. Steklov.* **38** (1951) 244–278. (Russian).

NOBLE, M. E.
(1) "Extensions and applications of a Tauberian theorem due to Valiron." *Proc. Cambridge Philos. Soc.* **47** (1951) 22–37.
(2) "Non-measurable interpolation sets. I. Integral functions." *Proc. Cambridge Philos. Soc.* **47** (1951) 713–732.
(3) "Non-measurable interpolation sets. II. Functions regular in an angle." *Proc. Cambridge Philos. Soc.* **47** (1951) 733–740.
(4) "Non-measurable interpolation sets. III. A theorem of B. J. Maitland." *Quart. J. Math. Oxford Ser.* (2) **4** (1953) 11–18.

OGURA, K.
(1) "Sur la théorie de l'interpolation de Stirling et les zéros des fonctions entières." *Bull. Sci. Math.* (2) **45** (1921) 31–40.

OKAMURA, H.
(1) "Sur la croissance des séries entières." *Mem. Coll. Sci. Kyota A* **19** (1936) 253–269.

PALEY, R. E. A. C., and WIENER, N.
(1) "Fourier Transforms in the Complex Domain." Colloquium Publ. No. 19, Amer. Math. Soc., Providence, Rhode Island, 1954.

PENNYCUICK, K.
(1) "On a theorem of Besicovitch." *J. London Math. Soc.* **10** (1935) 210–212.
(2) "Extension of a theorem of Faber–Pólya." *J. London Math. Soc.* **12** (1937) 267–272.

PFLUGER, A.
(1) "Zur Defektrelation ganzer Funktionen endlicher Ordnung." *Comment. Math. Helv.* **19** (1946) 91–104.

PHRAGMÉN, E.
(1) "Sur une extension d'un théorème classique dans la théorie des fonctions." *Acta Math.* **28** (1904).

PHRAGMÉN, E., and LINDELÖF, E.
(1) "Sur l'extension d'un principe classique de l'analyse et sur quelques propriétés des fonctions monogènes dans le voisinage d'un point singulier." *Acta Math.* **31** (1908) 381–406.

PICARD, M. E.
(1) "Sur une propriété des fonctions entières." *C. R. Acad. Sci. Paris* **88** (1879) 1024–1027.

PLANCHEREL, M., and PÓLYA, G.
(1) "Fonctions entières et intégrales de Fourier multiples." *Comment. Math. Helv.* **9** (1937) 224–248; **10** (1938) 110–163.

POISSON, S. D.
(1) "Mémoire sur le calcul numérique des intégrales définies." Memoires de l'Academie Royale des Sciences de l'Institut de France, vi (1823, published 1827), 571–602, particularly p. 575.

POLLARD, H.
(1) "Integral transforms." *Duke Math. J.* **13** (1946) 307–330.
(2) "Integral transforms, II." *Ann. Math.* (2) **49** (1948) 956–965.

PÓLYA, G.
(1) "Neuer Beweis für die Produktdarstellung der ganzen transzendenten Funktionen endlicher Ordnung." Sitzungsber Bayerl. Akad., 1921.
(2) "Über das Anwachsen von ganzen Funktionen, die einer Differentialgleichung genügen." *Vierteljahresschr. Naturforsch. Ges.* Zürich, 1916.
(3) "On an integral function of an integral function." *J. London Math. Soc.* **1** (1926) 12–15.
(4) "Über Annäherung durch Polynome mit lauter reellen Wurzeln." *Rend. Circ. Mat. Palermo* **36** (1913) 279–295.
(5) "Über die Nullstellen sukzessiver Derivierten." *Math. Z.* **12** (1922) 36–60.
(6) "Bemerkungen über unendliche Folgen und ganze Funktionen." *Math. Ann.* **88** (1923) 169–183.
(7) "Untersuchungen über Lucken und Singularitäten von Potenzreihnen." *Math. Z.* **29** (1929) 549–640.
(8) "On the zeros of the derivatives of a function and its analytic character." *Bull. Amer. Math. Soc.* **49** (1943) 178–191.
(9) "Remarks on characteristic functions." *Proc. Berkeley Symp. Math. Stat. Probability* (1945, 1946) Univ. of California Press, Berkeley, California, 1949, pp. 115–123.
(10) "On the zeros of an integral function represented by Fourier's integral." *Messenger Math.* **52** (1923) 185–188.
(11) "On the minimum modulus of integral functions." *J. London Math. Soc.* **1** (1926) 78–86.

PÓLYA, G., and SZEGÖ, G.
(1) "Aufgaben und Lehrsätze aus der Analysis," Vol. I, 2nd ed. Springer-Verlag, Berlin and New York, 1954.

PORTER, M. B.
(1) "On a theorem of Lucas." *Proc. Nat. Acad. Sci. U.S.A.* **2** (1916) 247–248, 335–336.

PRINGSHEIM, A.
(1) "Elementare Theorie der ganzer transzendenten Funktionen von endlicher Ordnung." *Math. Ann.* **58** (1904) 257–342.

RADEMACHER, H.
(1) "Über die asymptotische Verteilung gewisser konvergenzerzeugender Faktoren." *Math. Z.* **11** (1921) 276–288.

RADÓ, T.
(1) "Subharmonic Functions." Springer-Verlag, Berlin, 1937.

RAJAGOPAL, C. T.
(1) "On inequalities for analytic functions." *Amer. Math. Monthly* **60** (1953) 693–695.

RAUCH, A.

(1) "Extensions de théorèmes relatifs aux directions de Borel des fonctions méromorphes." *J. Math.* **12** (1933).

REDDY, A. R.

(1) "On the Maximum Modulus and the Mean Modulus of an Entire Function." *Canad. Math. Bull.* **12** No. 6 (1969) 869–872.

REDHEFFER, R. M.

(1) "Remarks on the incompleteness of $\{e^{i\lambda_n x}\}$, nonaveraging sets, and entire functions." *Proc. Amer. Math. Soc.* **2** (1951) 365–369.

(2) "On even entire functions with zeros having a density." *Trans. Amer. Math. Soc.* **77** (1954) 32–61.

(3) "On a theorem of Plancherel and Pólya." *Pacific J. Math.* **3** (1953) 823–835.

RÉNYI, A., and RÉNYI, C.

(1) "Some remarks on periodic entire functions." *J. Anal. Math.* **14** (1965) 303–310.

RIESZ, F.

(1) "Sur les fonctions subharmoniques et leur rapport à la théorie du potentiel." *Acta Math.* **54** (1930) 321–360.

RIESZ, F., and RIESZ, M.

(1) "Über die Randwerte einer analytischen Funktion." *C. R. 4e Congr. Math. Scand. Stockholm* (1916), 27–44.

RIESZ, M.

(1) "Sur le principe de Phragmén-Lindelöf." *Proc. Cambridge Philos. Soc.* **20** (1920) 205–207; and correction, *ibid.* **21** (1921) 6.

RITT, J. F.

(1) "On the iteration of rational functions." *Trans. Amer. Math. Soc.* **21** (1920) 348–356.

(2) "Prime and composite polynomials." *Trans. Amer. Math. Soc.* **23** (1922) 51–66.

ROSENBLOOM, P. C.

(1) "The fix-points of entire functions." *Medd. Lunds Univ. Mat. Sem. Tome Suppl.* (1952), 186–192.

ROTH, A.

(1) "Approximationseigenschaften und Strahlengrenzwerte meromorpher und ganzer Funktionen." *Comment. Math. Helv.* **11** (1938) 77–125.

SAKS, S., and ZYGMUND, A.

(1) "Analytic Functions." Monografie Matematyczne, Vol. 28. Warsaw, 1952.

SAXER, W.

(1) "Sur les valeurs exceptionelles des dérivées successives des fonctions méromorphes." *C. R. Acad. Sci. Paris* 182 (1926) 831–833.

SCHAEFFER, A. C.

(1) "Entire functions and trigonometric polynomials." *Duke Math. J.* **20** (1953) 77–88.

SCHOENBERG, I. J.

(1) "On certain two-point expansions of integral functions of exponential type." *Bull. Amer. Math. Soc.* **42** (1936) 284–288.

(2) "On Pólya frequency functions, I. The totally positive functions and their Laplace transforms." *J. Anal. Math.* **1** (1951) 331–374.
(3) "On Pólya frequency functions, II. Variation-diminishing operators of the convolution type." *Acta Sci. Math. Szeged* 12, Leopoldo Fejér et Frederico Riesz LXX annos natis dedicatus, Pars B (1950) 97–106.

SCHOTTKY, F.
(1) "Über den Picardschen Satz und die Borelschen Ungleichungen." *Sitzungsber. Preuss. Akad. Wiss.* (1904) 1244–1263.
(2) "Über zwei Beweise des allgemein Picardschen Satzes." *Sitzungsber. Kgl. Preuss. Akad. Wiss.* Berlin, 1907.

SEIDEL, W.
(1) "On the distribution of values of bounded analytic functions." *Trans. Amer. Math. Soc.* **36** (1934) 201–226.

SELBERG, A.
(1) "Über ganzwertige ganze transzendente Funktionen, I, II." *Arch. Math. Naturvid.* **44** (1941) 45–52, 171–181.
(2) "Über einen Satz von A. Gelfond." *Arch. Math. Naturvid.* **44** (1941) 159–170.
(3) "Eine Ungleichung der Potentialtheorie und ihre Anwendung in der Theorie der meromorphen Funktionen." *Comment. Math. Helv.* **18** (1945–65).
(4) "Über eine Ungleichung der Potentialtheorie." *Comment. Math. Helv.* **18** (1946).

SHAH, S. M.
(1) "A theorem on integral functions of integral order." *J. London Math. Soc.* **15** (1940) 23–31.
(2) "A theorem on integral functions of integral order, II." *J. Indian Math. Soc.* **5** (1941), 179–188.
(3) "A note on meromorphic functions." *Math. Student* **12** (1944) 67–70.
(4) "On integral functions of perfectly regular growth." *J. London Math. Soc.* **14** (1939) 293–302.
(5) "A note on the classification of integral functions." *Math. Student* **9** (1941) 63–67.
(6) "Note on a theorem of Pólya." *J. Indian Math. Soc.* **5** (1941) 189–191.
(7) "On integral functions of integral or zero order." *Bull. Amer. Math. Soc.* **48** (1942) 329–334.
(8) "The lower order of the zeros of an integral function." *J. Indian Math. Soc.* **6** (1942) 63–68.
(9) "The lower order of the zeros of an integral function, II." *Proc. Indian Acad. Sci. Sect. A.* **21** (1945) 162–174.
(10) "On the relations between the lower order and the exponent of convergence of zeros of an integral function." *J. Univ. Bombay* **11**, Part III (1942) 10–13.
(11) "The maximum term of an entire series." *Math. Student* **10** (1942) 80–82.
(12) "The maximum term of an entire series, II." *J. Indian Math. Soc.* **9** (1944) 54–55.
(13) "The maximum term of an entire series, III." *Quart. J. Math. Oxford Ser.* (1) **19** (1948) 220–223.
(14) "The maximum term of an entire series, IV." *Quart. J. Math. Oxford Ser.* (2) **1** (1950) 112–116.
(15) "The maximum term of an entire series, V." *J. Indian Math. Soc.* **13** (1949) 60–64.
(16) "The maximum term of an entire series, VI." *J. Indian Math. Soc.* **14** (1950) 21–28.
(17) "The maximum term of an entire series, VII." *Ganita* **1** (1950) 82–85.

(18) "A note on the maximum modulus of the derivative of an integral function." *J. Univ. Bombay* **13** Part III (1944) 1–3.
(19) "On proximate orders of integral functions." *Bull. Amer. Math. Soc.* **52** (1946) 326–328.
(20) "On the lower order of integral functions." *Bull. Amer. Math. Soc.* **52** (1946) 1046–1052.
(21) "A note on the minimum modulus of a class of integral functions." *Bull. Amer. Math. Soc.* **53** (1947) 524–529.
(22) "A note on the derivatives of integral functions." *Bull. Amer. Math. Soc.* **53** (1947) 1156–1163.
(23) "A note on lower proximate orders." *J. Indian Math. Soc.* **12** (1948) 31–32.
(24) "A note on uniqueness sets for entire functions." *Proc. Indian Acad. Sci. Sect. A* **28** (1948) 519–526.
(25) "On the coefficients of an entire series of finite order." *J. London Math. Soc.* **26** (1951) 45–46.
(26) "A note on entire functions of perfectly regular growth." *Math. Z.* **56** (1952) 254–267.

SHEFFER, I. M.
(1) "Concerning Appell sets and associated linear functional equations." *Duke Math. J.* **3** (1937) 593–609.
(2) "Some properties of polynomial sets of type zero." *Duke Math. J.* **5** (1939) 590–622.
(3) "Some applications of certain polynomial classes." *Bull. Amer. Math. Soc.* **47** (1941) 885–898.

SHIMIZU, T.
(1) "On the theory of meromorphic functions." *Japan. J. Math.* **6** (1929) 119–171.

SHTEINBERG, N. S.
(1) "On the interpolation of entire functions." *Mat. Sb.* **3** (72) (1952) 559–574 (Russian).

SIDDIQI, J. A.
(1) "Quelques théorèmes d'unicité." *C. R. Acad. Sci. Paris* **236** (1953) 1727–1729.

SIKKEMA, P. C.
(1) "Differential operators and differential equations of infinite order with constant coefficients." "Researches in Connection with Integral Functions of Finite Order." Noordhoff, Groningen-Djakarta, 1953.

SINGH, S. K.
(1) "A note on entire functions." *J. Univ. Bombay* **20**, Part 5, Sect. A (1952) 1–7.
(2) "A note on a paper of R. C. Buck." *Proc. Indian Acad. Sci. Sect. A* **38** (1953) 120–121.

SIRE, J.
(1) "Sur les fonctions entières de deux variables d'ordre apparent total fini. Thèse." *Rend. Circ. Mat. Palermo* **31** (1911) 1–91.
(2) "Sur la puisance de l'ensemble des points singuliers transcendants des fonctions inverses des fonctions entières." *Bull. Soc. Math. France* (1913).
(3) "Sur les fonctions entières de deux variables d'ordre apparent total fini et à croissance régulière." *J. Math.* (1913).

SPENCER, D. C.
(1) "Note on some function-theoretic identities." *J. London Math. Soc.* **15** (1940) 84–86.

SRIVASTAVA, P. L.
(1) "On two theorems of Akhyeser and a theorem of Cramér." *Rend. Circ. Mat. Palermo* (1) **55** (1931) 116–120.
(2) "A theorem on integral functions." *Bull. Acad. Sci. Allahabad* **1** (1932) 43–44, review by E. Ullrich, *Zentralblatt Math.* **6** (1933), 409.
(3) "On the Phragmén-Lindelöf principle." *Proc. Nat. Acad. Sci. India* (*Allahabad*) **6**, (1936) 241–243.

TEICHMÜLLER, O.
(1) "Vermutungen und Sätze über die Wertverteilung gebrochener Funktionen endlicher Ordnung." *Deutsch. Math.* **4** (1939) 163–190.

TIMAN, A. F.
(1) "On interference phenomena in the behavior of entire functions of finite degree." *Dokl. Akad. Nauk SSSR* **89** (1953) 17–20 (Russian).

TITCHMARSH, E. C.
(1) "The Theory of Functions." Oxford Univ. Press, London and New York, 1939.
(2) "On integral functions with real negative zeros." *Proc. London Math. Soc.* (2) **26** (1927) 185–200.
(3) "Introduction to the Theory of Fourier Integrals." Oxford Univ. Press, London and New York, 1937.
(4) "A theorem on infinite products." *J. London Math. Soc.* **1** (1926) 35–37.

TÖPFER, H.
(1) "Über die Iteration der ganzen transzendenten Funktionen, insbesondere von sin z und cos z." *Math. Ann.* **117** (1939) 65–84.

TRUTT, D.
(1) "Some mean squares of entire functions." Dissertation. Purdue University, Lafayette, Indiana, 1964.
(2) "Extremal norm-determining measures for Hilbert spaces of entire functions." *J. Math. Anal.* **20** (1967), 74–89.

TSUJI, M.
(1) "Wiman's theorem on integral functions of order $<\frac{1}{2}$." *Proc. Japan Acad.* **26** No. 2–5 (1950) 117–130.
(2) "Potential Theory in Modern Function Theory." Maruzen, Tokyo, 1959.
(3) "On the theorems of Carathéodory and Lindelöf in the theory of conformal representation." *Japan J. Math.* **7** (1930).
(4) "On the cluster set of a meromorphic function." *Proc. Imp. Acad. Japan* **19** (1943).
(5) "Theory of meromorphic functions in a neighbourhood of a closed set of capacity zero." *Japan J. Math.* **19** (1944).
(6) "Borel's directions of meromorphic functions of finite order, I." *Tohoku Math. J.* **2** (1950).
(7) "On the order of the derivative of a meromorphic function." *Tohoku Math. J.* **3** (1951).
(8) "On a direct transcendental singularity of an inverse function of a meromorphic function." *J. Math. Soc. Japan.* **5** (1953).
(9) "Theory of meromorphic functions on an open Riemann surface with null boundary." *Nagoya Math. J.* **6** (1953).

(10) "Borel's directions of a meromorphic function in a unit circle." *J. Math. Soc. Japan* **7** (1955).
(11) "On the cluster set of a meromorphic function." *Comment. Math. Univ. Sancti Pauli (Tokyo)* **4** (1955).

TUMURA, Y.
(1) "On the extensions of Borel's theorem and Saxer-Csillag's theorem." *Proc. Phys. Math. Soc. Japan* (3) **19** (1937) 29–35.

UHL, W.
(1) "Über die Darstellung ganzer Funktionen mittels der Stirling'schen Reihe bei Hermite'scher Interpolation." *Mitt. Math. Sem. Univ. Giessen* **33** (1944).

VALIRON, G.
(1) "Démonstration de l'existence pour les fonctions entières de chemins de détermination infinie." *C. R. Acad. Sci. Paris* **166** (1918) 382–384.
(2) "Les théorèmes généraux de M. Borel dans la théorie des fonctions entières." *Ann. Sci. École Norm. Sup.* (3) **37** (1920) 219–253.
(3) "Recherches sur le théorème de M. Picard." *Ann. Sci. École Norm. Sup.* (1921).
(4) "Sur les zéros des fonctions entières d'ordre fini." *Rend. Circ. Mat. Palermo* **43, 44** (1919, 1920).
(5) "Remarques sur le théorème de M. Picard." *Bull. Sci. Math. Ser. 2* **44** (1920) 91–104.
(6) "Sur les zéros des fonctions entières d'ordre infini." *C. R. Acad. Sci. Paris* **172** (1921) 741–744.
(7) "Recherches sur le théorème de M. Picard dans la théorie des fonctions entières." *Ann. Sci. École Norm. Sup.* (3) **38** (1922) 389–437; **39** (1922) 317–341.
(8) "Sur les fonctions entières vérifiant une classe d'équations différentielles." *Bull. Soc. Math. France* unpublished.
(9) "Sur les fonctions entières d'ordre entier." *C. R. Acad. Sci. Paris,* **174** (1922) 1054–1056.
(10) "Le théorème de Laguerre-Borel dans la théorie des fonctions entières." *Bull. Sci. Math. Ser. 2* **46** (1922) 432–445.
(11) Thèse, Toulouse (1914).
(12) "Theory of Integral Functions." Chelsea, Bronx, New York, 1949.
(13) "Sur les valeurs asymptotiques de quelques fonctions méromorphes." *Rend. Circ. Mat. Palermo* **49** (1925) 415–521.
(14) "Valeurs exceptionelles et valeurs déficientes des fonctions méromorphes." *C. R. Acad. Sci. Paris* **225** (1947) 556–558.
(15) "Sur la distribution des valeurs des fonctions méromorphes." *Acta Math.* **47** (1925) 117–142.
(16) "Sur les valeurs déficientes des fonctions méromorphes d'ordre nul." *C. R. Acad. Sci. Paris* **230** (1950) 40–42.
(17) "Fonctions Entières d'Ordre Fini et Fonctions Méromorphes." L'Enseignement Mathématique, Geneva (1960).
(18) "Sur les fonctions entières d'ordre fini." *Bull. Sci. Math. France* (2) **45** (1921) 258–270.
(19) "A propos d'un memoire de M. Pólya." *Bull. Sci. Math. France* (2) **48** (1924) 9–12.
(20) "Fonctions entières et fonctions méromorphes d'une variable." "Mèmorial des Sciences Mathématiques," No. 2. Gauthier-Villars, Paris, 1925.

(21) "Sur la formule d'interpolation de Lagrange." *Bull. Sci. Math. France* (2) **49** (1925) 181–192, 203–204.

(22) "Sur un théorème de M. Wiman." *Opuscula Math. A. Wiman Dedicata, Lund* (1930) 1–12.

(23) "Sur une classe de fonctions entières admettant deux directions de Borel d'ordre divergent. *Compositio Math.* **1** (1934) 193–206.

(24) "Sur le minimum du module des fonctions entières d'order inférieur a un." *Math. Cluj* **11** (1935) 264–269.

(25) "Directions de Borel des fonctions méromorphes." "Mémorial des Sciences Mathématiques," No. 89. Gauthier-Villars, Paris, 1938.

(26) "Sur les fonctions entières d'ordre nul." *Math. Ann.* **70** (1911) 471–498.

(27) "Sur les fonctions entières d'ordre fini et d'ordre nul et en particulier les fonctions a correspondance régulière." (Thèse). *Ann. Fac. Sci. Univ. Toulouse* (3) **5** (1913) 117–257.

(28) "Sur quelques théorèmes de M. Borel." *Bull. Soc. Math. France* (1914).

(29) "Sur la croissance du module maximum des séries entières." *Bull. Soc. Math. France* **44** (1916) 45–64.

(30) "Sur les chemins de détermination des fonctions entières." *Bull. Soc. Math. France* (1917).

(31) "Sur les fonctions entières et leurs fonctions inverses." *C. R. Acad. Sci. Paris* **173** (1921) 1059–1061.

(32) "Sur un théorème de M. Fatou." *Bull. Sci. Math.* (2) **46** (1922) 200–208.

(33) "Recherches sur le théorème de M. Borel dans la théorie des fonctions méromorphes d'ordre fini." *Acta Math.* **52** (1928).

(34) "Sur les directions de Borel des fonctions méromorphes d'ordre fini." *J. Math. Série* **10** (1931).

(35) "Sur la derivée angulaire dans la représentation conforme." *Bull. Sci. Math.* **56** (1932) 208–211.

(36) "Points de Picard et points de Borel des fonctions méromorphes dans un cercle." *Bull. Sci. Math.* **56** (1932) 10–32.

VAN DER CORPUT, J. G., and SCHAAKE, G.

(1) "Ungleichungen für Polynome und trigonometrische Polynome." *Compositio Math.* **2** (1935) 321–361.

VIDENSKII, V. S.

(1) "Consequences of a proposition of S. N. Bernstein on entire functions of genus zero." *Dokl. Akad. Nauk SSSR* **84** (1952) 421–424 (Russian).

WAHLUND, A.

(1) "Über einen Zusammenhang zwischen dem Maximalbetrage der ganzen Funktion und seiner unteren Grenze nach dem Jensenschen Theoreme." *Ark. Mat.* **21A** (1929).

WALSH, J. L.

(1) "The location of critical points of analytic and harmonic functions." *Amer. Math. Soc. Colloq. Publ.* **34** (1950) 219.

(2) "On the location of the roots of the jacobian of two binary forms and of the derivative of a rational function." *Trans. Amer. Math. Soc.* **24** (1921) 106–116.

WEIERSTRASS, K.
(1) "Zur Theorie der eindeutigen analytischen Funktionen." Werke, Band II. Mayer and Müller, Berlin, 1895, pp. 77–124: reprinted from *Abhandl. Königl. Akad. Wiss.* (1876) 11–60.

WHITTAKER, E., and WATSON, G.,
(1) "A course of Modern Analysis," 4th ed. Cambridge Univ. Press, London and New York, 1935.

WHITTAKER, J. M.
(1) "The lower order of integral functions." *J. London Math. Soc.* **8** (1933) 20–27.
(2) "The order of the derivative of a meromorphic function." *J. London Math. Soc.* **11** (1936) 82–87.
(3) "Interpolatory function theory." "Cambridge Tracts in Mathematics and Mathematical Physics," No. 33. Cambridge Univ. Press, London and New York, 1935.

WHITTINGTON, J. E.
(1) "On the fix points of entire functions." *Proc. London Math. Soc.* (3) **17** (1967) 530–546.

WIDDER, D.
(1) "The Laplace transform." Princeton Univ. Press, London and New York, 1946.

WIGERT, S.
(1) "Sur un théorème concernant les fonctions entières." *Ark. Mat. Astr. Fys.* **11** (1916) No. 22.

WILSON, R.
(1) "Directions of strongest growth of the product of integral functions of finite order and mean type." *J. London Math. Soc.* **28** (1953) 185–193.

WIMAN, A.
(1) "Über eine Eigenschaft der ganzen Funktionen von der Höhe Null." *Math. Ann.* **76** (1914) 197–211.
(2) "Über die angenäherte Darstellung von ganzen Funktionen." *Ark. Mat. Ast. Fys.* **1** (1903).
(3) "Sur une extension d'un théorème de M. Hadamard." *Ark. Math. Astr. Fys.* **2** (1905).
(4) "Sur le cas d'exception dans la théorie des fonctions entières." *Arkiv. Math.* **1** (1952).
(5) "Über den Zusammenhang zwischen dem Maximalbetrage einer analytischen Funktion und dem grössten Gliede der zugehörigen Taylorschen Reihe." *Acta Math.* **37, 41**.

WISHARD, A.
(1) "Functions of bounded type." *Duke Math. J.* **9** (1942) 663–676.

WITTICH, H.
(1) "Neuere Untersuchungen über eindeutige analytische Funktionen." Ergeb. Math., No. 8. Springer-Verlag Berlin, 1955.

WOLF, F.
(1) "An extension of the Phragmén-Lindelöf Theorem." *J. London Math. Soc.* **14** (1939) 208–216.

INDEX

Pure and Applied Mathematics

A Series of Monographs and Textbooks

1: Arnold Sommerfeld. Partial Differential Equations in Physics. 1949 (Lectures on Theoretical Physics, Volume VI)
2: Reinhold Baer. Linear Algebra and Projective Geometry. 1952
3: Herbert Busemann and Paul Kelly. Projective Geometry and Projective Metrics. 1953
4: Stefan Bergman and M. Schiffer. Kernel Functions and Elliptic Differential Equations in Mathematical Physics. 1953
5: Ralph Philip Boas, Jr. Entire Functions. 1954
6: Herbert Busemann. The Geometry of Geodesics. 1955
7: Claude Chevalley. Fundamental Concepts of Algebra. 1956
8: Sze-Tsen Hu. Homotopy Theory. 1959
9: A. M. Ostrowski. Solution of Equations and Systems of Equations. Third Edition, in preparation
10: J. Dieudonné. Treatise on Analysis: Volume I, Foundations of Modern Analysis, enlarged and corrected printing, 1969. Volume II, 1970. Volume III, 1972
11: S. I. Goldberg. Curvature and Homology. 1962.
12: Sigurdur Helgason. Differential Geometry and Symmetric Spaces. 1962
13: T. H. Hildebrandt. Introduction to the Theory of Integration. 1963.
14: Shreeram Abhyankar. Local Analytic Geometry. 1964
15: Richard L. Bishop and Richard J. Crittenden. Geometry of Manifolds. 1964
16: Steven A. Gaal. Point Set Topology. 1964
17: Barry Mitchell. Theory of Categories. 1965
18: Anthony P. Morse. A Theory of Sets. 1965
19: Gustave Choquet. Topology. 1966
20: Z. I. Borevich and I. R. Shafarevich. Number Theory. 1966
21: José Luis Massera and Juan Jorge Schaffer. Linear Differential Equations and Function Spaces. 1966
22: Richard D. Schafer. An Introduction to Nonassociative Alegbras. 1966
23: Martin Eichler. Introduction to the Theory of Algebraic Numbers and Functions. 1966
24: Shreeram Abhyankar. Resolution of Singularities of Embedded Algebraic Surfaces. 1966
25: François Treves. Topological Vector Spaces, Distributions, and Kernels. 1967
26: Peter D. Lax and Ralph S. Phillips. Scattering Theory. 1967.
27: Oystein Ore. The Four Color Problem. 1967
28: Maurice Heins. Complex Function Theory. 1968

29: R. M. Blumenthal and R. K. Getoor. Markov Processes and Potential Theory. 1968

30: L. J. Mordell. Diophantine Equations. 1969

31: J. Barkley Rosser. Simplified Independence Proofs: Boolean Valued Models of Set Theory. 1969

32: William F. Donoghue, Jr. Distributions and Fourier Transforms. 1969

33: Marston Morse and Stewart S. Cairns. Critical Point Theory in Global Analysis and Differential Topology. 1969

34: Edwin Weiss. Cohomology of Groups. 1969

35: Hans Freudenthal and H. De Vries. Linear Lie Groups. 1969

36: Laszlo Fuchs. Infinite Abelian Groups: Volume I, 1970. Volume II, 1973

37: Keio Nagami. Dimension Theory. 1970

38: Peter L. Duren. Theory of H^p Spaces. 1970

39: Bodo Pareigis. Categories and Functors. 1970

40: Paul L. Butzer and Rolf J. Nessel. Fourier Analysis and Approximation: Volume 1, One-Dimensional Theory. 1971

41: Eduard Prugovečki. Quantum Mechanics in Hilbert Space. 1971

42: D. V. Widder: An Introduction to Transform Theory. 1971

43: Max D. Larsen and Paul J. McCarthy. Multiplicative Theory of Ideals. 1971

44: Ernst-August Behrens. Ring Theory. 1972

45: Morris Newman. Integral Matrices. 1972

46: Glen E. Bredon. Introduction to Compact Transformation Groups. 1972

47: Werner Greub, Stephen Halperin, and Ray Vanstone. Connections, Curvature, and Cohomology: Volume I, De Rham Cohomology of Manifolds and Vector Bundles, 1972. Volume II, Lie Groups, Principal Bundles, and Characteristic Classes, in preparation

48: Xia Dao-xing. Measure and Integration Theory of Infinite-Dimensional Spaces: Abstract Harmonic Analysis. 1972

49: Ronald G. Douglas. Banach Algebra Techniques in Operator Theory. 1972

50: Willard Miller, Jr. Symmetry Groups and Their Applications. 1972

51: Arthur A. Sagle and Ralph E. Walde. Introduction to Lie Groups and Lie Algebras. 1973

52: T. Benny Rushing. Topological Embeddings. 1973

53: James W. Vick. Homology Theory: An Introduction to Algebraic Topology. 1973

54: E. R. Kolchin. Differential Algebra and Algebraic Groups. 1973

55: Gerald J. Janusz. Algebraic Number Fields. 1973

56: A. S. B. Holland. Introduction to the Theory of Entire Functions. 1973

In preparation

Samuel Eilenberg. Automata, Languages, and Machines: Volume A

H. M. Edwards. Riemann's Zeta Function

Wayne Roberts and Dale Varberg. Convex Functions